Dr. Geo. Vasey

The agricultural Grasses and forage Plants of the United States

Dr. Geo. Vasey

The agricultural Grasses and forage Plants of the United States

ISBN/EAN: 9783337140120

Printed in Europe, USA, Canada, Australia, Japan

Cover: Foto ©berggeist007 / pixelio.de

More available books at **www.hansebooks.com**

J. S. DEPARTMENT OF AGRICULTURE.

BOTANICAL DIVISION.

SPECIAL BULLETIN.

THE

AGRICULTURAL GRASSES AND FORAGE PLANTS

OF THE •

UNITED STATES;

AND SUCH FOREIGN KINDS AS HAVE BEEN INTRODUCED.

By Dr. GEO. VASEY, (Botanist;

WITH AN

APPENDIX

On the CHEMICAL COMPOSITION OF GRASSES, by CLIFFORD
RICHARDSON, and a glossary of terms used
in describing grasses.

A NEW, REVISED, AND ENLARGED EDITION,
WITH 114 PLATES. •

PUBLISHED BY AUTHORITY OF THE SECRETARY OF AGRICULTURE.

WASHINGTON:
GOVERNMENT PRINTING OFFICE.
1889.

LETTER OF SUBMITTAL.

SIR: Herewith I present a Report on the Agricultural Grasses and Forage Plants of the United States, with illustrations.

This report is largely a revised and enlarged edition of the "Agricult. ural Grasses of the United States" published by this Department in 1884. In the present report the principal forage plants, other than grasses, which are employed in agriculture, are treated of. The purpose is to give plain, general descriptions of the plants, together with practical notes and observations from persons who have tested them in actual cultivation or who have given them special investigation. The larger portion of the plants treated of are natives of the United States, but such foreign species as have been tried here have also received proper notice.

The great extent of this country, with its extraordinary diversities of climate and soil, makes necessary a corresponding diversity in the subjects and methods of agriculture. With respect particularly to grasses and forage plants adapted to different sections of the country we are yet in the infancy of our knowledge, and must patiently and intelligently conduct such experiments as will give us the precise information we need. Every farmer and stock-raiser in the country is interested in this subject, and it has been the endeavor of the writer to present to the attention of such persons in any part of the country some grasses or forage plants suitable to their wants.

Respectfully,

GEO. VASEY,
Chief of Botanical Division.

Hon. J. M. RUSK,
Secretary of Agriculture.

3

AGRICULTURAL GRASSES AND FORAGE PLANTS OF THE UNITED STATES.

Every thoughtful farmer realizes the importance of the production on his land of a good supply of grass for pasturage and hay. He who can produce the greatest yield on a given number of acres will be the most successful man; yet this is a subject which has been, and still is, greatly neglected.

In the United States we have many climates, many kinds of soil, many geological formations, many degrees of aridity and moisture. It must be apparent that one species of grass can not be equally well adapted to growth in all parts of this extensive territory; yet hardly a dozen species of grasses have been successfully introduced into our agriculture. True it is that this number answers with a tolerable degree of satisfaction the wants of quite an extensive portion of the country, chiefly the northern and cooler regions. But it is well known that in other localities the same kinds of grasses do not succeed equally well, and one of the most important problems for those regions is to obtain such kinds as shall be thoroughly adapted to their peculiarities of climate and soil. This is particularly the case in the Southern and Southwestern States, the arid districts of the West, and in California.

The solution of this question is largely a matter of experiment and observation.

The grasses which we have in cultivation were once wild grasses, and are still such in their native homes.

The question then arises, can we not select from our wild or native species some kinds which will be adapted to cultivation in those portions of the country which are not yet provided with suitable kinds? Many observations and some experiments in this direction have already been made, and if proper research is continued, and sufficiently thorough experiments are followed up, there is no reason to doubt that proper kinds will be found for successful cultivation in all parts of the country.

The plains lying west of the one hundredth meridian, together with much broken and mountainous interior country, nearly treeless and arid, in New Mexico, western Texas, and Arizona, are unreliable for the purposes of ordinary agriculture, but are becoming more and more

5

important as the great feeding ground for the multitudes of cattle which supply the wants of the settled regions of our country as well as the constantly increasing foreign demand. The pasturage of this region consists essentially of native grasses, some of which have acquired a wide reputation for their rich nutritious properties, for their ability to withstand the dry seasons, and for the quality of self-drying or curing, so as to be available for pasturage in the winter. This quality is due probably to the nature of the grasses themselves and to the effect of the arid climate. It is well known that in moist countries, at lower altitudes, the grasses have much succulence; they grow rapidly, and their tissues are soft; a severe frost checks or kills their growth, and chemical changes immediately occur which result in rapid decay; whereas in the arid climate of the plains the grasses have much less succulence, the foliage being more rigid and dry, and therefore when their growth is arrested by frost the tissues are not engorged with water, the dessicating influence of the climate prevents decay, and the grass is kept on the ground in good condition for winter forage. General Benjamin Alvord, of the U. S. Army, in an article on the subject of these winter-cured grasses, states that they only acquire this property on land which is 3,000 feet above the level of the sea. The region having such an altitude includes, he says, all, nearly up to the timber line, of Montana, Idaho, Wyoming, Utah, Nevada, Colorado, and New Mexico; five-sixths of Arizona, one-half of Dakota, one-fourth of Texas, one-fifth of Kansas, and one-sixth each of California, Oregon, and Washington Territory, embracing about one-fourth of the area of the whole United States.

Many of the grasses of this extensive region are popularly known as "bunch grass," from their habit of growth; others are known as "mesquite" and "grama grass." These consist of many species of different genera, some of them more or less local and sparingly distributed, others having a wide range from Mexico to British America.

The most important of the "bunch grasses" may be briefly mentioned as follows: Of the genus *Stipa* there are several species; *Stipa comata* and *Stipa setigera* occur abundantly in New Mexico, Texas, Arizona, and California, reaching to Oregon. In Colorado, Kansas, and all the prairie region northward, stretching into British America, *Stipa spartea* is the principal one of the genus. On the higher plateaus and near the mountains the *Stipa viridula* is very common, extending from Arizona to Oregon and British America. Somewhat related botanically is *Oryzopsis cuspidata*, a very rigid bunch grass, with a fine, handsome panicle of flowers. It is equally wide-spread with the preceding. Another widely diffused grass is *Deschampsia cæspitosa*, varying much in size and thriftiness according to the altitude and amount of moisture where it grows, but always having a light, elegant, spreading panicle of silvery gray flowers.

One of the most extensively diffused grasses is *Kœleria cristata*, varying in height from 1 foot to 2½ feet, with a narrow and closely flowered spike. Several species of fescue grass (*Festuca*) are intermixed with the vegetation in varying proportions; the most important of these probably are *Festuca ovina* in several varieties, and *Festuca scabrella*, the latter especially in California, Oregon, and Washington.

The genus *Calamagrostis* (or *Deyeuxia*, as it has been called) furnishes several species which contribute largely to the vegetation of this region. They are mostly tall, stiff, and coarse grasses, but leafy and some of them very nutritious. Of these, *Calamagrostis sylvatica* and *Calamagrostis neglecta* are the least valuable. Perhaps the best of them is *Calamagrostis Canadensis*, which is soft and leafy. Next in value, probably, is *Calamagrostis Aleutica*, of California and Oregon, extending into Alaska. *Calamagrostis (Ammophila) longifolia*, confined chiefly to the plains east of the Rocky Mountains, is tall and reed-like, growing in dense clumps, from 4 to 6 feet high.

Several species of *Andropogon* are diffused from Arizona to British America, but are not found on the western coast. The principal species are *Andropogon scoparius, A. furcatus*, and *A. (Chrysopogon) nutans*. Some of them are known under the name of "bluejoint."

Other grasses also widely spread, but in more sparing quantity, are several species of *Poa* and *Glyceria*. Several varieties of *Agropyrum repens*, or couch grass, occur abundantly in saline soils, and also *Agropyrum glaucum*, which is widely known as "blue stem," and is considered among the most nutritious of grasses. *Brizopyrum spicatum*, now called *Distichlis maritima*, and some species of *Sporobolus*, also form extensive patches or meadows in saline soils. Besides there is a large number of grasses of low growth and of more spreading habit, which are known in the southwest and east of the Rocky Mountains under the names of "mesquite" and "buffalo" grasses. The former belong mostly to the genus *Bouteloua*, the most important species being *B. racemosa*, or tall mesquite, and *B. oligostachya*, or low mesquite. The true buffalo grass is, botanically, *Buchloë dactyloides*, which in many places forms extensive fields over large areas. It is of a low and densely tufted or matted habit. Another similar grass, but of little value, spreading out in low, wide patches, is *Munroa squarrosa*. The above-mentioned species form the larger proportion of the grassy vegetation of the great plains.

GRASSES FOR GENERAL CULTURE.

The grasses form one of the largest and most widely diffused families of plants, being spread over all habitable parts of the globe. Some kinds are restricted to particular localities, others are diffused over large

forms on account of the size, quantity, and nutritive value of their grains. The fact of their great value being discovered, the observation would soon follow that by planting the seeds in suitable ground, and caring for the growing plants by the exclusion of all other vegetation, a certain and reliable resource for sustenance would be obtained.

This was the beginning of agriculture, and agriculture made possible the numerical increase and diffusion of human population.

History of Grass Culture.—The selection and cultivation of particular kinds of grasses with reference to their superior grazing qualities and for the greater production of hay is, however, a comparatively modern practice.

In the Philippine Islands, as we are informed by the United States consul at Manilla (Mr. Julius G. Voight), a species of rice grass (*Leersia hexandra*) is cultivated for the purpose of supplying feed for the few domestic animals which are kept for the cultivation of land and for the carrying of burdens.

9

This grass (locally called *zacate*) is cultivated exclusively in low, wet ground, and is flooded occasionally after the manner of rice, being first started in seed beds and then transplanted to the previously flowed field. How far this custom prevails in other eastern countries we do not know, but from the general antiquity and uniformity of the practices of husbandry in those countries we may suppose that this practice is there of ancient origin.

But as far as western nations are concerned the cultivation of special grasses for hay is a modern improvement. Mr. Martin J. Sutton, in a recent work on "Permanent and Temporary Pastures," states that *Lolium perenne*, or perennial rye grass, was the first grass gathered separately for agricultural purposes. He further states that it has been known since 1611, the date of the earliest agricultural book which mentions it. Mr. George Sinclair, in his advertisement to the fourth edition of the "Hortus Gramineus Woburnensis," says:

The time has been in this country [*i. e.*, England] when providing sufficient forage for live stock in winter was a matter of the greatest difficulty, and great losses were sustained, and many advantages given up, on account of the absolute want of winter fodder. Old turf, suitable either for grazing or for the scythe, was supposed to be a *creation of centuries*, and that a farmer, who wished to lay down a meadow in his youth, must see the end of his "three score years and ten" before he could possibly possess a piece of pasture capable of keeping a score of sheep or a couple of cows. So much was the want of grass land felt among arable farmers in times past that the tenancy of it was eagerly sought, its value was consequently highly prized, and heavy fines were imposed for breaking it up. The banks of rivers were usually made commonable, in order that the surrounding farmers might each have a share; and these meadows were in many cases irrigated in order to increase still more the scanty stock of winter fodder.

Perennial rye grass, as we have seen, began to be cultivated early in the seventh century, and it seems to have been about the only grass so cultivated for a hundred years longer. In 1763 it is said that a Mr. Wynch brought from Virginia into England the *Phleum pratense*, under the local name of Timothy grass, it having been cultivated in the United States for some forty years. This was also soon established as an agricultural grass in England, and a few years later was followed by the introduction of orchard grass (*Dactylis glomerata*) from Virginia, by the Society of Arts; at least this statement is made by Mr. Parnell in his work on British grasses, but is probably an error. It is considered doubtful by Mr. Charles Johnson in the "Grasses of Great Britain," who says it is eminently European, being distributed naturally over the whole of Europe and the adjoining parts of Asia. It is not known to be native in the United States.

As to *Phleum pratense* (Timothy grass), it is naturally widely diffused over Europe, but it is admitted by all that its cultivation was first undertaken in the United States, where it is also indigenous in mountainous regions. It is, however, well known that in Europe up to about the year 1815 there were but three or four kinds of grass generally cultivated.

At that time the Duke of Bedford instituted his famous series of experiments at Woburn, in England, for determining the nutritive properties of different grasses. These experiments brought into notice many before unnoticed grasses and greatly stimulated their cultivation; and the subsequent development of this branch of agriculture has been the means of obtaining astonishing results, not only in the multiplied facilities for the grazing and fattening of cattle and sheep, but also in the reaction of this business on the cultivation of grain, by the greatly multiplied means of obtaining manures by which the exhausted lands were renewed and the yield of grain increased.

History of Grass Culture in the United States.—In the early history of this country, particularly in the Northern States, while the settlements were sparse, the natural pasturage was abundant, and the natural meadows and marshes furnished a supply of hay for winter feeding. But in course of time, by the increase of population, the farms began to crowd each other, and the range for cattle was restricted.

Then probably arose the question of forming meadows and pastures of limited extent. Early in the last century Mr. Jared Elliot (of Connecticut) made some valuable investigations respecting the grasses suitable for cultivation, and by practice and teaching sought to bring this subject to the attention of the people.

In 1749 he wrote a particular account of the fowl meadow grass (*Poa serotina*) which is native in New England, giving an interesting account of its value as a meadow grass.

He also refers to Herd's grass, or Timothy, as having been found "in a swamp in Piscataqua by one Herd, who propagated the same." It is also said to have been cultivated in Maryland about the year 1720. This was some fifty years before its cultivation in England. It is also stated by Parnell in his work on the British Grasses, that orchard grass (*Dactylis glomerata*) was first cultivated in the United States, and thence introduced into England about the middle of the eighteenth century. Probably soon after this date two other standard grasses came into use, viz, *Poa pratensis* (Kentucky blue grass) and *Agrostis alba* (redtop). Some other grasses have had a limited trial, but the Timothy grass, blue grass, orchard grass, and redtop have continued to be the principal meadow grasses of the Northern States. To these should be added red clover, which, although not a grass, is a very common meadow crop, usually combined with Timothy.

Grass in the South.—Although the Southern States were earlier settled than the Northern ones, there was a very different condition of agriculture as respects grazing and hay-making. In some of these States the climate permits of the growth of grasses during the greater part of the year, some species making their growth during the hot season and others during the colder months, so that cattle may commonly obtain subsistence in the field throughout the year, and hay is little employed except for horses and cattle kept to labor.

But these places suffer from protracted droughts in summer and fall, which parch the pastures so that cattle and sheep are not then able to find a sufficiency of feed. The pasture and meadow grasses of the North have not been generally cultivated with success in the States which border on the Gulf of Mexico, and the greatest want of agriculture in that region is the introduction of grasses that will maintain growth and vigor during protracted droughts.

The same remarks may be made with respect to the grasses needed for cultivation in the arid districts of the West, and there is every reason to expect that grasses adapted to such conditions of climate and soil will be found. .

Permanence of Pastures and Meadows.—It has long been a question as to how long land should be allowed to continue in pasture or meadow. The answer to this question will depend very much on circumstances.

Unquestionably the best plan for farming is the practice of mixed husbandry, or a mixture of raising grain crops and the fattening of domestic animals; for with a diversity of products there is an alleviation of the evils of frequent crop failures, which are usually limited to one or two kinds, and also an alleviation of the fluctuations in the prices of crops, so that where some grain crops fail from any cause, the farmer has a resource in those of another kind and in his live stock. Besides, the rotation of crops, including the periodical laying down of cultivated ground to grass, and the change of grass land to the growth of field crops, results in the best condition of the soil.

In the practice of most farmers, meadow lands are seldom continued more than three or four years without a change to the plow. But pasture lands are more frequently kept undisturbed for a longer time, and so long as they continue in a healthy, clean, and productive state there can be no objection to their permanence; but whenever a pasture becomes overgrown with weeds, or filled with worthless or unproductive grasses, it is time for it to take its place in a system of rotation and renovation, at the same time regarding the needs of the soil in respect to fertilizing and cleaning from rocks, briers, and other shrubs.

Drainage of Grass Lands.—Generally speaking, there is the same benefit to be derived from the proper drainage of grass lands, that is so conspicuously shown in lands devoted to other crops. All lands with an impervious subsoil of stiff clay, or soils that are water-clogged, may be greatly benefited by proper draining, both in the quality and quantity of the grass product. On such land, properly drained, the grass will start earlier in the spring and will continue to grow later in the fall than without drainage. All soils which rest upon a porous subsoil do not need it, and land may have so strong a slope that the water is discharged from it with sufficient rapidity without the aid of a drain. Wet, water-soaked pastures generally abound in rushes and sedges, which may grow luxuriantly, but are coarse and innutritious. The valuable grasses on such pastures are injured or destroyed by the tramping of cattle, whose hoofs penetrate the wet ground.

An eminent German scientist has demonstrated that there is an intimate connection between a warm, dry soil and economy in feeding cattle. Friable land absorbs more heat than land which is saturated with moisture, and retains the heat for a longer period. Upon the one, animals lie warmer, especially at night, than they do upon the other. Now a large portion of the food consumed by animals is utilized for the production of the heat which is constantly dissipated from their bodies. It follows that additional food becomes necessary to replace the animal heat lost by the colder surroundings.*

The Selection of Grasses.—The selection of the proper kinds of grasses to be employed for meadows or pastures must depend on several circumstances, such as soil, drainage, habit of growth, productions, etc. No one kind of grass can be expected to be adapted to all conditions, neither can any given mixture of grasses. There has been a great amount of empiricism in this matter. One man finds a certain grass to be very thrifty and productive on his farm, and thinks he has found the great desideratum, and at once proclaims his grass, perhaps gives it a new name, and recommends its use, without regard to the conditions or circumstances which may be absolutely essential to its success.

Others purchase seed of the new grass, perhaps at exorbitant prices, and without a knowledge of its peculiar habits or wants, give it a trial and find it a failure, probably because climate or soil, or other essential conditions are unsuitable to its wants.

Mr. Sutton, writing on this subject, says:

The whole question is one of experience, and I am well persuaded that those who possess the largest knowledge, drawn from the widest sources, will concur in the opinion that each individual case should be considered independently and upon its own merits. I would lay great stress upon the necessity of starting with a clear understanding of the condition and capability of the soil. The subsoil, too, must be taken into account; for sooner or later its influence will tell decisively upon the existence of certain grasses.

Then the purpose of the grass crop must not be overlooked. Whether it is chiefly for hay or entirely for grazing will prove an important consideration in determining the sorts to be sown. Even the kind of cattle the land is intended to carry is worth more than a passing thought. Milch cows, fattening stock, sheep, and horses, or a combination of these animals, can be provided for if a definite object is held steadily in view.

In an old and well-settled country there is much accumulated experience among farmers, which a beginner may avail himself of to the avoidance of serious mistakes. Still an observing and progressive man will often find occasion for a departure from established rules and practices in the introduction of new kinds for cultivation; indeed it is only thus that progress and improvement can be made; but it will also be wise to make such experiments with caution and without incurring too much risk.

In some portions of our country the experience of the past is very unsatisfactory with respect to grass culture; and in other portions, as

* Sutton on Permanent and Temporary Pastures.

in the new settlements of the arid districts, all culture must be in the nature of experiment, and much judgment and large information are needed to guide the experimenter to the best results.

Relation of Stock to Pastures.—The farmer and grazier should always bear in mind that his pastures should be adapted to the kind as well as the quantity of stock which he keeps. Cattle and sheep are very different in their feeding habits, the sheep cropping the grass very close, and cattle requiring to have the grass longer in order to get a bite. Horses again do not bite as close as cattle. By judiciously proportioning the kind of stock kept on the pasture a much better result may be obtained by keeping both cattle and sheep than by keeping either alone. The field will thus be kept cleaner and in better condition.

Management of the Pasture.—Care must be observed that cattle or sheep be not put upon grass too early in the spring, before the grass has fairly commenced to grow. This rule applies particularly to sheep, who will in such cases eat the heart out of the grass crown, to its entire destruction. When, however, the grasses have made a good start there will be much of the taller stalks and coarser culms which the sheep will reject, and which cattle will crop with avidity. As the season advances there are often bunches of grass neglected by both cattle and sheep, giving to the pasture a rough and uneven appearance, when the mower should be run over the pasture, after which the old tufts will send up another crop of tender blades.

No precise date can be given for beginning to graze pastures in the spring. Cattle should not be turned in until there is enough feed to keep them going without too much help from hay, nor until the ground is firm enough to prevent their hoofs from damaging the young shoots of the grasses.

On the other hand, if the grass gets too old, the animals refuse much of it, and the fodder will be lost. Pastures consisting largely of early, strong-growing grasses, particularly cock's foot (orchard grass), will need to be stocked before others which produce finer and later varieties.*

It is sometimes a nice question to determine when to take stock off the pastures in the fall. This will depend much on the length of the growing season in any particular locality. In northern latitudes the growth of vegetation will be arrested early, and when the grass has quite ceased to grow the stock should be removed that the ground may be in proper condition for an early start in the following spring. Usually, however, in northern sections of the country the question is effectually settled by the early descent of the winter snows. In southern latitudes the climate is so mild that the growing season continues all winter, so that stock live mainly or entirely upon the growing grass, there being sorts there which naturally make their principal growth in the coolest portion of the year.

---

* Sutton on Permanent and Temporary Pastures.

Supplementary Feed.—It often happens that a drought occurs in the summer or fall, in which the pastures are dried and parched so that the cattle fail to get a sufficient amount of feed. It is, therefore, the practice of careful and provident farmers to have a tract of land sown to some kind of fodder, which may be drawn upon to supply the deficiency of pasturage, and not only to keep the animals from suffering, but to keep them also in a growing condition. Corn sown broadcast or in close drills, or sorghum sown in like manner, are some of the best grasses for this purpose.

Some varieties of sweet corn, combining earliness and productiveness or large size, will be better than common field corn, especially to keep up the supply of milk from cows.

Hungarian grass and millet make excellent fodder crops. They are both considered to be but varieties of the same species, and there is practically little difference between them. If sowed on tolerably rich ground they will produce sometimes a very large yield of grass. They are of rapid growth, and are frequently ready to be cut two months from the time of sowing. They generally produce an abundance of nutritious seeds, on account of which cattle thrive better on them than on corn fodder. Beets and prickly comfrey are also recommended as fodder plants in some localities.

The pastures may also often be relieved by turning stock on to stubble after harvest.

Humanity dictates that a man should not keep any more stock than he can under ordinary circumstances care for and give sufficient feed. But a provident and good manager will be enabled safely to keep a much larger number than a man who is shiftless and careless. He will do this by making provision for casualties and probable contingencies. It is much better and more profitable to have a surplus of feed than to have a deficiency.

Kind of Grasses for Meadows and Pastures.—In this country there has been very little variety in the kinds of grasses cultivated, the range being generally Timothy, blue grass or June grass, orchard grass, and redtop, usually combined more or less with red or white clover.

Farmers are influenced somewhat by the markets they supply. The most popular hay in the markets of the great cities is Timothy, and meadows of this grass alone are very common, and when well managed are very satisfactory and profitable. It is also very common to combine Timothy with red clover in various proportions.

In low, wet meadows, particularly in New England, redtop is considerably employed, and it is a common constituent of pastures in all the Northern States.

In England, great attention has been given to combining several kinds of grasses in meadows, and it is claimed that the practice is better for the land and gives a larger yield than when one variety only is employed. By using a mixture the ground may often be more uni-

formly covered, and in pastures there will be, from the different flowering time of the different species, a succession and continuation of a
supply of tender foliage.

Some species of grass are adapted to clay lands, some to sandy soils,
some to loam, some to dry upland, and some to low land; but even for
land of a uniform quality it is believed that a mixture of five or six suitable varieties will yield a larger crop than one alone. The mixture of
several varieties is perhaps most valuable in land that is designated for
pasturage, as then they reach maturity at different times and furnish
a succession of good feed, and also cover more completely and uniformly
the ground. But no general mixture of grass seed can be adapted to all
situations and soils. Every farmer should study carefully the nature
of his grounds, its altitude, drainage and composition, and then adapt
his grasses to the circumstances.

Generally there are few cases where there will be any advantage in
employing more than five or six well-selected varieties for cultivation
in one field. For a permanent pasture under most circumstances the
following kinds in proper proportions would make a good mixture, viz:
June grass (blue grass), fox-tail (*Alopecurus pratensis*), redtop (bent
grass), Timothy, tall fescue, and perennial rye grass. This will give
a succession as to earliness of growth and flowering.

But in some localities and for some soils, as in Kentucky for instance,
the farmer who has a good pasture of blue grass will not think it capable of much improvement. As we speak of the individual kinds of
grasses and their adaptation to different soils, the farmer will be able
to judge how far they will suit his circumstances.

Mixed Grasses for Pasturage.—For pasturage, however, we recommend a variety of grasses and thick seeding. Stock like variety and thrive better on it. Each
variety has its season of greatest excellence, and thus the best pasturage can be kept
up throughout the year. The common red clover should be sown with the grasses
for all pastures. It is a rank grower and resists drought admirably. We are glad
more attention is being paid to pasturage. Improved farming can not be carried on
without it, and in nothing are the majority of our farmers more neglectful than in
seeding more of their farms to good pastures.*

A Kentucky farmer gives the following mixture where an immediate
pasture is wanted:

Blue grass ....pounds..	8
Orchard grass ....do....	4
Timothy ....do....	4
Red clover ....do....	6

To this may be added Italian rye grass, 4 pounds, and the same
amount of fescue grass if preferred, but the other is ordinarily sufficient.
This quantity is a heavy seeding for one acre. The blue grass will
not be seen much at first, but by the time the clover dies out it will
have taken hold of the entire surface.

A writer in the New England Farmer recommends the following formula for a permanent pasture:

Early varieties—
Red clover ....pounds.. 10
Alsike clover ....do.... 5
Orchard grass ....bushel.. 1
June grass ....do.... 1
Perennial rye grass ....do.... 1
Late varieties—
Herd's grass ....do.... ½
R. I. bent grass...do.... ½
Redtop ....do.... 1

This forms an unusually heavy seeding, and probably the quantities may be advantageously reduced, but the combination presents a variety that will give a succession from early till late in the season.

The more common mixture for meadows is as follows per acre:

Redtop ....bushel.. 1
Timothy ....do.... ½
Red clover...pounds.. 4

On highlands orchard grass might be substituted for the redtop.

Time and Manner of Seeding Grass Seed.—There has been much diversity of opinion as to the proper time of seeding land to grass. A very common practice has been to sow the seed in the spring with a grain crop, generally of oats. If the season is favorable this method succeeds very well, having the advantage of no loss in the regular crops of the land. The growing grain furnishes to the young grass shelter and shade from the heat of the sun, and after the removal of the crop the grass spreads, and sometimes the same season furnishes a light crop for the scythe or some grazing for the cattle. But the success of this plan of seeding is not by any means certain. In a very dry season the young plants may perish from drought, or in a wet season the grain may lodge and smother the young grass. Hence others recommend late summer or early fall seeding. A writer in the Massachusetts Ploughman makes the following statement:

The last half of August is generally considered the best time for seeding; earlier than this the weather is apt to be too hot for the ready germination of the seed, and weeds will get a start before the grass. The first half of September is a good time, and we have sometimes had very good success with seeding as late as October 1, but would prefer to sow earlier if possible. If rye is sown with the grass seed it is best done about the middle of September; too much rye will choke the grass, but a light seeding of about one-half to five-eighths of a bushel per acre will not injure the grass much, and will give a much better return the next season than the grass alone.

Too little care is usually bestowed upon the preparation of the land for seeding; it should be worked only when just moist enough to make the lumps crush easily, and should be harrowed repeatedly and rolled before sowing the seed, then brushed and rolled again, which will leave the land in fine, smooth order for the mowing-machine or scythe.

It is customary to mix Herd's grass, redtop, and clover seed in seeding, but we prefer to seed high land with Herd's grass (*Phleum pratense*) only low, moist land with redtop (*Agrostis vulgaris*) and fescue, and clover by itself in the spring, for the reason that the season of maturity of these grasses is very different; the clover should be cut about the 15th of June while in blossom, the Herd's grass about July 1, and the redtop about July 15. When they are mixed it will be impossible to cut them all in perfection; and if the Herd's grass is cut too early in dry weather it is almost sure to be killed out.

Mr. T. C. Alvord, of Vermont, writes in the Boston Cultivator as follows:

For a number of years past I have sown grass seed only in the spring. On such lands as I wish to seed down without grain I fit my land in the fall if I can, as that saves valuable time in the spring; but if I do not have time to perform the work in the fall, I fit the land as early as I can in the spring, sowing the seed then. On all lands that I seed down I finish working the land before the seed is sown, never covering the seed. I think where grass seed is harrowed, raked, or brushed in much of the seed is covered so deep that it never comes up.

Many people think that grass seed sown in the spring will not make a good crop of hay the first season, and that it requires two seasons to do it. This is an error. On all the lands that I have sown with grass seed in the spring the grass has been ripe enough to cut in from ten to twelve weeks from the time the seed was sown, while I invariably get better crops than I do when I seed down with grain. If the grain lodges it will kill the grass, and if the weather is dry the grass will dry up, while in both cases the land will need reseeding; also weeds and foul grasses will occupy the soil.

If grass seed is sown by itself in the spring it will generally get so good a start that no ordinary dry or hot weather in the summer will injure the crop. When seeding land in this way a sufficient quantity of seed should be sown, so that if it all grows the land will be all occupied with grass, thus preventing the growth of weeds, also giving a large yield with a better quality of grass, while forming a thicker turf to be turned under for the enrichment of the soil when the land is again plowed.

On all lands which I have seeded in this way the first crop of hay has averaged two tons per acre.

Reseeding Old and Worn-out Meadows.—We have already stated that all wet lands with a clay subsoil should be subjected to a system of tile drainage, but in some cases a temporary substitute may be found in a certain manner of plowing, as is detailed in the following communication from a correspondent of the American Cultivator:

I will state my experience, in brief, on cold, wet, swale-land that was once a black-ash swamp. The grass was so light and wild it did not pay for cutting. Immediately after haying I plowed it in deep wide furrows, being sure to lap them and turn flat over. I took pains to make dead furrows where they should be, and also a clear outlet at the lower end of the furrows. I harrowed lightly with a fine harrow, and went over the field with a hoe and fixed the loose sod, and top-dressed with a light coat of manure and gravelly loam scraped up in the milking yard, and sowed on a mixture of redtop, timothy, and English flat turnip seed, then brushed lightly. Now for results: In the first place, I harvested turnips enough from the piece to pay for the labor of plowing and fitting the piece, and the next harvest I cut the heaviest burden of hay from that land that I ever saw on any meadow; it was waist-high and very thick. I accounted for it in this way, the land was thoroughly drained by the spaces left between the furrows, and the decaying sod provided a rich, warm seed-bed above the cold, wet, hard-pan, a portion of which had been brought to the surface by the deep plowing. Of course a roller would not have been tolerated on the piece, as it would have been detrimental to the best results. I wanted to get the land up and keep it up as long as I could, and let it breathe by leaving space for air to pass in under and come up through; and I believe that if such land was plowed in that way clear up to freezing time and seeded then or left until early spring, when clover seed could be added, most excellent results would follow.

GENERAL DESCRIPTION OF GRASSES.

A grass possesses the following parts : (1) The root, (2) the stem, (3) the leaves, (4) the flowers.

(1) The roots are the fibrous branching organs which extend downward into the ground and appropriate the water or other liquid nutriment to be conveyed into the stem and leaves.

(2) A stem that rises above ground, either erect, ascending, or reclining, is called a culm. In some species, in addition to the culm, there are horizontal subterranean stems, improperly called roots. They are known botanically as rhizomes, and are sometimes several feet long. They may be distinguished from the true roots by their bearing a greater or less number of scales and sending out erect branches as well as fibrous roots. In some grasses there is a kind of bulb at the base of the stem, in which is stored a concentrated mass of food for the support of the plant under peculiar circumstances, as in protracted drought. This bulbous formation is a part of the stem, and not of the root. The stem or culm of grasses is usually cylindrical and hollow; sometimes it is more or less compressed or flattened. It is divided at intervals by transverse thickened portions called joints or nodes, at which points leaves and sometimes branches are given off. These nodes tend also to strengthen the stem. Stems are usually simple and unbranched, except at the top, where they commonly divide into the more or less numerous branches of the panicle or flowering part. But some stems give rise from the side joints to leafy branches, which may, like the main stem, produce smaller panicles at the top.

(3) The leaves take their origin at the nodes or joints in two ranks— that is, they are placed alternately on opposite sides of the stem at greater or less distances; thus, the first leaf will be on one side, the second on the opposite side a little higher up, the third still higher and directly over the first, the fourth over the second, and so on. The leaves consist of three parts : (1) the sheath, (2) the ligule, and (3) the blade. The sheath is that part which clasps the stem. It is generally open on one side, as will be readily observed in the leaves of a corn-stalk, but in some grasses the sheath is partly or even completely closed together by the adhesion of the opposite edges. The sheath is analogous to the stem or petiole of the leaves of many higher plants. At the point where the blade of the leaf leaves the stem, at the top of the sheath and o·

19

the inner side, there is usually a small, thin, membranous organ, called the ligule or tongue. This is sometimes half an inch long, more commonly only two or three lines, and sometimes it is almost absent or reduced to a short ring, but its length and size are very constant in the same species. This ligule represents the stipules which occur at the base of the leaves in many of the higher plants. The blade or lamina is the expanded part of the leaf, but is usually called by the general name leaf. In the majority of grasses the leaf is long and narrow; that is, many times longer than wide. There is one central nerve, called the midnerve or midrib, extending to the point of the leaf, with numerous finer nerves on each side running parallel to it, and not connected by conspicuous transverse nerves nor giving off branches. These leaves are in some species rough, in others smooth, hairy, or downy, etc. The agricultural value of a grass depends mainly upon the quantity, quality, size, and nutritive properties of the leaves.

(4) The flowers of the grasses are generally at the end of the stem or the side branches, sometimes very few in number, sometimes in great abundance, sometimes in a close spike, and sometimes in a panicle, with many spreading branches or rays. The flowers may be single on the branches or on the pedicels, or they may be variously clustered. In the common redtop (*Agrostis alba* or *A. vulgaris*) there is a single flower at the end of each of the small branchlets of the panicle. Each of these flowers is inclosed by a pair of small leaf-like scales or chaff, called the outer or empty glumes. The flower consists of (1) the essential organs and (2) the envelopes. The essential organs are the stamens and pistils, which may readily be seen when the grass is in bloom. The stamens, of which there are usually three in each flower, consist of the anther and filament, the anther being the small organ which contains the pollen or dust which fertilizes the pistil or female organ, and the filament being the thread-like stem on which the anther is borne.

The pistil is the central organ of the flower, and consists of three parts; the ovary, the style, and the stigmas. In most grasses the style is divided into two branches. The stigmas are the delicate organs, usually of a plumose form, at the extremities of these branches, which receive the pollen for the fertilization of the flower; and the ovary is that part at the base which contains the future seed.

The envelopes of the flower are usually two leaf-like scales or husks, inclosing between them the stamens and pistil. These scales face each other, one being a very little higher on the axis than the other, and also usually smaller and more delicate in texture. This smaller scale is called the palet; the other larger and usually coarser one the flowering glume; its edges generally overlap and partly inclose the palet.

The flower constituted as above described, together with the pair of outer or empty glumes at the base, form what is called a spikelet. In many cases, however, there are two, three, or more flowers, sometimes even ten to twenty, in one spikelet. in which case they are arranged alternately on opposite sides of the axis, one above the other, with a pair of empty or outer glumes at the base of the cluster. Such may be seen in the blue grass (*Poa pratensis*), fescue grass (*Festuca*) and many others.

There are innumerable modifications of these floral organs, and upon the differences which exist in them the distinctions of genera and species are based. In some cases the glumes are entire in outline, in some they are toothed and lobed, and sometimes running out into a slender point called an awn, sometimes with a bristle or awn on the back, etc. They also vary in size from the twentieth part of an inch to an inch or more in length.

PLATE 1.

FIG. 1. 1, fibrous roots; 4, culm; 5, node; 6, leaf.

2. 2, rhizoma; 4, culm; 6, blade of leaf; 7, ligule; 9, scales of the rhizoma.

3. 1, root fibers; 3, bulbous base of culm; 4, culm; 5, sheath; 6, blade.

4. 2, scaly rhizomas; 4, node; 6, blade; 7, ligule; 9, scales of the rhizoma.

5. 1, fibrous roots; 2, creeping rhizoma; 4, culm; 5, sheath; 6, blade; 7, culm; 8, nodes.

PLATE 2.

The numbers in each of the figures are as follows: 1, sheath; 2, blade; 3, culm; 4, node, or joint; 5, ligule.

The ligule is best shown in the lower right-hand figure.

PLATE 3.

FIG. 1. A dense spike (Alopecurus pratensis).

2. An elongated, one-sided spike (Paspalum dilatatum).

3. Spike (Hordeum pratense).

4. Spike (Agropyrum repens).

5. Spike (Elymus condensatus).

6. Spike (Bouteloua polystachya).

7. Spike (Bouteloua oligostachya).

8. Panicle (Panicum Crus-galli).

9. Panicle (Agrostis exarata).

10. Panicle (Kœleria cristata).

11. Panicle (Distichlis maritima).

12. Panicle (Bromus secalinus).

13. Panicle (Hierochloa borealis).

14. Panicle (Poa pratensis).

15. Panicle (Dactylis glomerata).

PLATE 4.

FIG. 1. Two spikelets, one closed, one opened, of Agrostis vulgaris.

2. Two spikelets, one closed, one opened, of Agrostis exarata.

3. Two spikelets, one closed, one opened, of Sporobolus Indicus.

4. An opened spikelet of Calamagrostis Canadensis.

5. Two spikelets, one closed, one opened, of Phleum pratense.

6. Two spikelets, one closed, one opened, of Muhlenbergia diffusa.

7. Two spikelets, one closed, one opened, of Paspalum dilatatum.

8. Two spikelets, one closed, one opened, of Paspalum læve.

9. A spikelet of Aristida purpurea.

10. Two spikelets, one closed, one opened, of Setaria setosa.

Fig. 11. Two spikelets, one closed, one opened, of Setaria glauca.

12. Two spikelets, one closed, one opened, of Alopecurus pratensis.

13. Two spikelets, one closed, one opened, of Holcus lanatus.

14. A spikelet of Deschampsia cæspitosa and one of its flowers.

15. A spikelet of Poa serotina and one of its flowers.

16. A spikelet of Bromus erectus and one of its flowers.

17. The male and female spikelets of Buchloë dactyloides, the former both closed and opened.

PASPALUM.

In this genus the panicle does not divide into numerous slender branches as in many other kinds, but the flowers are arranged in several rows on one side of a narrow, flattened branch, called a rhachis. Each flower consists of two empty glumes of equal or nearly equal length, of a flowering glume of a thickish, hard texture, the edges of which overlap a palet of similar texture, and between these two are inclosed the stamens and pistils.

This genus has its range principally in the Southern and Southwestern States. The species are very numerous, are mostly perennial, and vary much in form and habit. Some are tall and erect, some decumbent or spreading, and others have the habit of sending out runners, which take root at short intervals and thus spread and form dense patches. They are all relished by cattle, and some of them are considered valuable as pasture grasses.

Paspalum dilatatum.

This may be called the hairy-flowered Paspalum. It has been found native in Virginia, Tennessee, Alabama, Mississippi, Louisiana, and Texas, and has been introduced into other States. It also occurs in South America. It grows from 2 to 5 feet high, with numerous leaves about a foot in length and one-third to one half an inch in breadth. It does not creep upon the ground like the following species, but is inclined to grow in tufts, which may attain considerable size. It is recommended both for pasture and hay by the few who have tried it.

This species has also been called *Paspalum ovatum*, but the name above given, having been first applied, is the proper one.

Charles N. Ely, Smith Point, southeastern Texas, says:

Paspalum dilatatum was brought to this country about twelve years ago, and planted by S. B. Wallis. It is a promising grass for hay and pasture, growing best on moist lands, but doing well on upland. It is easily subdued by cultivation, and is not inclined to encroach on cultivated lands. It is best propagated by roots or sets, the seed not being reliable. It is rather slow in starting, but when well rooted it spreads and overcomes all other grasses. Tramping and grazing is more of an advantage to it than otherwise. I think that this grass will succeed in a great variety of soils and climates, but those planting it must have patience with it at first.

Mr. Wallis, above referred to, says:

This I consider the most valuable of all the grasses with which I am acquainted; it is perennial and grows here all the year round, furnishing excellent green feed for stock at all seasons, except that the green blades freeze in our coldest weather perhaps two or three times in a winter. It increases rapidly from seeds, and also reproduces itself from suckers, which sprout from the nodes of the culm after the first crop of seed has ripened. I have seen these suckers remain green for six or eight weeks after the old stalks were as dead and dry as hay, and then, when the old stalks had fallen to the ground, take root and form new plants. It grows well on all kinds of dry land. Plants two or three years old form stools 12 to 18 inches across. The grass has very strong roots and grows in the longest drought almost as fast as when it rains.

(Plate 5.)

Paspalum platycaule.

This has sometimes been called Louisiana grass. It occurs in all the Gulf States and in the West Indies and South America. It grows flat on the ground rooting, at every joint, and forming at the South a thick, permanent, evergreen sod. It does well on almost any upland soil, and is said to stand drought better than Bermuda grass. It usually grows too short and close to the ground for hay, but for grazing it apparently has many good properties. It may be distinguished from the other Paspalums and from Bermuda grass by its flattened stems (whence the name) and the very slender seed-stalks, each bearing only two or three very narrow, somewhat upright spikes. The leaves, especially on the long runners, are short and blunt.

The facts of its being a perennial and seeding freely, of its doing better than any other grass on poor soil, forming a compact tuft to the exclusion of other plants, and of its being easily killed by cultivation, will doubtless recommend it for more extended growth.

Dr. Charles Mohr, Mobile, Ala., says:

It has taken a firm foot-hold in this section. It is perfectly hardy, prefers damp localities, and suffers somewhat from long droughts. It grows best in a sandy loam, rather close, compact, and damp, in exposed situations, as it does not stand shade well. It stands browsing and tramping well, and is greedily eaten by all kinds of stock. Its vegetation begins earlier in spring than that of Bermuda.

G. A. Frierson, Frierson's Mill, La., in the Southern Live Stock Journal, says:

It grows everywhere in rather low, wet, clay lands, and stands grazing as well or better than Bermuda.

B. H. Brodnax, Morehouse Parish, La.:

Paspalum platycaule was first noticed here about 1870 in very small patches. Since then it has spread rapidly from seed. It is not cultivated. It stands frost very well when firmly rooted, staying green nearly all winter, and it stands drought splendidly. It grows best on a poor quality of land high above overflow, or where water could not stand on it. It is a splendid pasture grass, making a sod equal to Bermuda, but it is not cut for hay. It is very easily destroyed, one plowing being sufficient to kill it.

Mr. Prentice Bailey, of Baker County, northern Florida, sends a specimen of Paspalum platycaule for identification, and says of it:

On all old roads, where travel has killed the other grasses and packed the soil, it covers the ground with a close, even turf; it forms such a thick turf that it is called here "blanket grass." The cattle in the woods are so fond of it and keep it eaten down so close, that it is difficult to find any of it more than 2 or 3 inches in height, but on good ground in protected places it grows to the height of several feet. It is only partially killed through the winter. From the avidity with which it is eaten by all kinds of stock, the closeness of turf formed, its ability to resist almost any amount of trampling, and its rapidity of growth I think that it is a most valuable grass for this country.

Mr. F. W. Thurow, of Harris County, Tex., says that at present *Paspalum platycaule* furnishes about five-eighths of the pasturage in southeastern Texas, forming a dense sod. Stock of all kinds seem to relish it, but is not as nutritious as Bermuda grass. (Plate 6.)

Paspalum distichum.

Several species of Paspalum have received attention in the South as being useful pasture grasses and very durable from their creeping and rooting habit. *Paspalum distichum* is one of these species. It grows principally in low, moist ground. Its stems and culms are mostly prostrate and running, sending up here and there a few flower-bearing culms. It is found in the Southern States and Texas, thence to California. Farther south it is found in most tropical countries. Mr. W. A. Sanders, of Fresno County, Cal., writes recently as follows :

Are you aware of the value of *Paspalum distichum* for seeding pond-holes that dry up or nearly so in autumn ? Such ponds are usually spots of bare, stinking mud, but when well set to this grass will yield all the way up to 80 tons (in the green state) of autumn feed for stock, especially valuable for cows first, then follow with sheep till every vestige is devoured. Surely it has an immense food value in such places.

(Plate 7.)

BECKMANNIA.

Beckmannia erucæformis (Slough Grass).

This genus is closely related to Panicum and has considerable resemblance to some forms of *Panicum Crus-galli*. It grows abundantly in the Rocky Mountain Region from California and Oregon eastward as far as Iowa and Minnesota. It is found in marshy ground and in sloughs, particularly in the neighborhood of streams.

It usually grows in tufts, and is of a coarse growth, the stout, roughish culms rising to about 3 feet in height; the thickish leaves are about half an inch wide and 6 to 8 inches long. These, as well as the loose, long sheaths, are strongly marked with numerous parallel veins. The panicle is generally long and narrow, from 6 to 10 inches long, and half an inch to an inch wide, composed mostly of many very short, closely-set branches, which are more or less interrupted below where the branches are generally longer, sometimes 2 inches long and erect.

The spikelets are crowded very closely together on the one-sided spikes, and each one consists of a pair of thickish, compressed, inflated, boat-shaped, empty glumes, and between these, one lanceolate, acute flowering glume, of thinner texture, with its still thinner palet, and the stamens and styles. These are represented in plate 8, *a* showing an enlarged spikelet, *b* the same expanded to show the separate parts. In some localities this grass is abundant and forms a valuable resource for stock. The bottom leaves and sterile shoots are tender and much relished. (Plate 8.)

PANICUM.

In this genus the mode of inflorescence is very variable, but most of the species have a spreading, much-branched panicle, the terminal branchlets of which have spikelets of a single perfect flower, or in some cases with a lower male or imperfect flower also. There are two or three empty glumes, the lower one generally much shorter than the others; the perfect flower has a thick, hard glume with a palet similar in texture, and with the stamens and pistil inclosed between them. The other imperfect flower when present has a glume similar to the empty ones.

The name is probably derived from the Latin word *panis*, bread, because some of the species were used, and are still used, for bread-making. The species of this genus are very numerous (more than three hundred on the globe), and of widely different appearance. We have about fifty native species, most of which have little practical value except as adding more or less to the wild forage of our woods and fields. But some species, both native and foreign, are of very high agricultural value.

Panicum maximum (True Guinea Grass).

This is a native of Africa, which has been introduced into many tropical countries, and in the West Indies is extensively cultivated. It has been brought into Florida and other places along the Gulf coast, but is little known in the United States. It requires a long season, is very susceptible to frost, and ripens seed only in the warmest part of the country. It has often been confounded with Johnson grass, which is very different. A sufficient point of distinction is the fact that Johnson grass spreads by underground stems, while Guinea grass does not, but remains in bunches.

Its chief value is for hay or soiling, and it should be cut frequently to prevent it becoming too hard and coarse. It grows tall and rank, reaches the height of 6 or 8 feet when mature, and yields a seed much resembling millet. It is not adapted to the climate of the Northern States. *Panicum jumentorum* is a synonym. (Plate 9.)

Panicum Texanum (Texas Millet).

This grass is a native of Texas, and was first described and named in 1866 by Prof. S. B. Buckley, in his preliminary report of the "Geographical and Agricultural Survey of Texas." It is frequently called Colorado grass, from its abundance along the Colorado River in that State. In some localities it is known as river grass; in others as goose grass, from its being supposed to have been introduced by wild geese. In southern Texas it is sometimes called buffalo grass, and in Fayette County it is known as Austin grass from the fact that it was first utilized as hay near Austin.

The most numerous and favorable reports regarding it are from Lampasas, Burnet, and Travis Counties, along the Colorado River, and southward through the central part of the State. From no grass, so little known, have more favorable reports been received, especially from the section in which it is most abundant. It is but little known outside of Texas. Of the thirty five valuable reports in regard to it, all but six were from that State, and most of them from the region above indicated.

The grass is an annual, growing usually from 2 to 4 feet high, and is especially valuable for hay. It prefers rich alluvial soils, but stands drought well, though on dry uplands its yield is much reduced. The plant is furnished with an abundance of rather short and broad leaves, and the stems, which are rather weak, are often produced in considerable number from a single root, and where the growth is rank are inclined to be decumbent at the base. It is valuable for all purposes for which the ordinary millets are used, and should be tried throughout the South. In Texas, where most largely grown, it generally overcomes other grasses and weeds; but in some of the other Southern States crab grass and weeds have interfered with its growth. It has not been much cultivated in the Northern States, but is deserving of a trial; as with a good season it will probably be more productive than, and of superior quality to, common Hungarian millet. (Plate 10.)

Panicum proliferum, var. geniculatum.

This variety occurs in the Southern States, where it is sometimes called "sprouting crab grass." It is an annual, growing in low, moist ground. The stems are first erect, then become decumbent and spreading, frequently attaining a length of 6 or 7 feet, bent and rooting at the lower joints. It has much the same habit as *P. Texanum*, but the stems are smooth and more flattened; the leaves also are smoother and longer. The stems are sometimes nearly an inch thick at the base and very succulent. The main stem is terminated by a diffuse panicle sometimes 2 feet long.

Dr. Charles Mohr, of Mobile, says of it:

In damp, grassy places it prefers rich ground throughout the coast region. It commences to vegetate vigorously in the hottest part of the summer, throwing out numerous shoots from the joints, forming large-branched bushes. The foliage is rich and tender; and the succulent, thick stems are sweet and juicy. After cutting, it throws out numerous sprouts from the lower joints, which grow rapidly, so as to allow repeated cuttings until frost. It is through all stages of its growth much relished by horses and cattle.

(Plate 11.)

Panicum barbinode (Para Grass).

This species has been introduced from South America in some localities of the Southern States. In Cuba it is cultivated and highly valued for its prolific growth and nutritive properties. It is not adapted to culture in the Northern States. It is a coarse, reed-like grass, that looks as if it should grow in the water; but it makes a heavy growth on the high pine ridges of Florida. (Plate 12.)

Panicum miliaceum.

This is the millet grass of India, or at least one of the Indian millets. It has, in Asia, been cultivated for ages, and is, in many parts, an important article in the food supply of the natives. It is also cultivated in Egypt, Turkey, and Southern Europe. It has been cultivated to a limited extent in this country for forage, and will thrive and ripen in the Northern as well as the Southern States.

Mr. Charles L. Flint says:

Millet is one of the best crops we have for cutting and feeding green for soiling -purposes, since its yield is large, its luxuriant leaves juicy and tender and much relished by milch cows and other stock. The seed is rich in nutritive qualities, but it is very seldom ground or used for flour, though it is said to exceed all other kinds of meal or flour in nutritive elements. An acre well cultivated will yield from 60 to 70 bushels of seed. Cut in the blossoms, as it should be for feeding to cattle, the seed is comparatively valueless. If allowed to ripen its seed, the stalk is no more nutritious, probably, than oat straw. It is well adapted to culture in dry regions.

(Plate 13.)

Panicum Crus-galli (Barnyard Grass).

This is an annual grass, with thick, stout culms usually from 2 to 4 feet high. In the Southern States it is often employed, and is consid ered a valuable grass. Professor Phares, of Mississippi, says:

In that and some other States it is mowed annually, and is said sometimes to furnish four or five tons of hay per acre. It annually reseeds the ground and requires no cultivation or other care, save protection from stock and the labor of harvesting. In one county in Mississippi hundreds of acres are annually mowed on single farms. Cows and horses are very fond of it whether green or dry.

In the Northern States it is seldom employed. (Plate 14.)

Panicum sanguinale (Crab Grass).

This is an annual grass, which, although a native of the Old World, has become spread over most parts of this country, and indeed over all tropical countries. It is the most common crab grass of the Southern States. It occurs in cultivated and waste grounds, and grows very rapidly during the hot summer months. The culms usually rise to the height of 2 or 3 feet, and at the summit have from three to six slender flower spikes, each from 4 to 6 inches long. The culms are bent at the lower joints, where they frequently take root. At the New Orleans Exposition there were specimens of this grass 5 feet 10 inches long.

Professor Killebrew, of Tennessee, says:

It is a fine pasture grass; although it has but few base leaves and forms no sward, yet it sends out numerous stems or branches at the base. It serves a most useful purpose in stock husbandry. It fills all our corn-fields and many persons pull it out, which is a tedious process. It makes a sweet hay, and horses are exceedingly fond of it, leaving the best hay to eat it.

Professor Phares, of Mississippi, says that the corn and cotton fields are often so overrun with it that the hay which might be secured would be more valuable than the original crop. It is sometimes mowed from between the rows, sometimes cut across the ridges, with the corn.

Although so much esteemed in the South, it is considered a pest in the Northern States. (Plate 15.)

Panicum virgatum (Tall Panic Grass; Switch Grass).

A tall perennial grass, 3 to 5 feet high, growing mostly in clumps in moist or even in dry, sandy soil, very common on the sea-coast, and also in the interior to the base of the Rocky Mountains. This is a good and prolific grass if cut when young; when ripe it becomes harsh and unpalatable. It forms a large constituent of the native grasses of the prairies, particularly in moist localities. It is said to be cultivated in some parts of Colorado, and with very satisfactory results. (Plate 16.)

Panicum agrostoides. (Redtop Panicum.)

This is a perennial grass, commonly growing in large clumps in wet meadows or on the muddy margins of lakes and rivers. It grows 4 to 6 feet high, is erect in habit, and developes its reddish panicles from several of the joints as well as at the apex. The stem is somewhat flattened and very smooth, as are the sheaths; the leaves are 1 to 2 feet long, about half an inch wide, and somewhat rough on the margins and midrib. The terminal panicle is 6 to 12 inches long, at first somewhat close, but becoming quite open and diffuse. The lateral panicles are shorter and partly inclosed by the sheath at the base. The branches of the panicle are mostly 1 or 2 inches long and rather densely flowered nearly to the base. The spikelets are a little more than a line long on very short pedicels, mostly racemose on one side of the branches, oblong, acute, the lower empty glume ovate, acute, half as long as the upper one, which is rather long-pointed and five-nerved; the lower or sterile flower is a little shorter than the longer glume and a little shorter than the perfect flower, which is oblong, obtuse, and under a lens shows a few beards at the apex. This grass produces a large amount of foliage, which makes fair hay if cut before flowering time; if left later it contains too many wiry stalks. It may be utilized as a hay crop in low grounds, but it is doubtful if it can be made productive on dry, tillable land. (Plate 17.)

Panicum anceps. (Two-edged Panic Grass.)

A perennial grass, when well developed resembling the preceding, but of a smaller, lighter growth, generally found in moist clay soil. It has a flattish erect stem, 2 to 3 feet high, with smooth leaves a foot or more long, of a bluish-green color, and mostly near the base of the stem. The root-stock is thick, scaly, and creeping near the surface of the ground. The panicle is 6 to 12 inches long, with short branches near the top, the lateral branches 3 to 6 inches long, rather distant, erect or somewhat spreading. Usually there are also several smaller lateral panicles from the upper joints of the culm. The spikelets are about a line and a half long, a little longer than those of *Panicum agrostoides*, oblong, lanceolate, a little curved, and sessile, or on very short pedicels. The lower empty glume is broadly ovate, and about half as long as the five to seven-nerved upper one. The lower glume of the sterile flower is as long as the upper empty glume, and much like it in texture, while the palet is thin, obtuse, and much shorter. The perfect flower is one third shorter than the upper empty glume, oblong; the flowering glume and its palet, as in most species of *Panicum*, is thick and of hard texture. This can not be considered a valuable grass, but it frequently occurs in neglected and poor land in sufficient quantity to afford considerable grazing for stock. It makes its growth late in the season, usually reaching the flowering state in August. Dr. Mohr, of Mobile, remarks that it is not much relished by stock, being rather harsh and dry.

Professor Phares says:

It forms strongly rooted spreading clumps, often completely carpeting the ground with pretty, glossy, light-green foliage.

(Plate 18.)

SETARIA.

In this genus the flowers are constructed as in the Panicums, but they are arranged in narrow, more or less cylindrical spikes. Below the spikelets are several bristles, generally longer than the spikelets, which remain on the spike after the fall of the flowers.

Setaria Italica (Hungarian Grass; German Millet).

This grass is supposed to be a native of the East Indies, but it has been extensively introduced into most civilized countries. It has long been cultivated as a fodder-grass both in Europe and in this country. It is an annual grass of strong, rank growth, the culms erect, 2 to 3 feet high, with numerous long and broad leaves, and a terminal, spike-like, nodding panicle, 4 to 6 inches long, and often an inch or more in diameter. The varieties and forms of this grass differ greatly, so much so that some of them have been considered different species; but the general opinion of botanists is that they are all varying forms of the same species, dependent upon the character of the soil, thickness of seeding, moisture or dryness, and time of sowing. It owes its value as a fodder plant to the abundance of its foliage, and to the large quantity of seed produced. In some instances objection has been made to this grass on account of the bristles which surround the seed, and which have been said to penetrate the stomachs of cattle so as to cause inflammation and death. But it is plain that this opinion is not generally held, as the cultivation of the grass is widely extended and everywhere recommended.

For forage it should be cut as soon as it blooms, when, of course, it is worth nothing for seed; but it is most valuable for forage and exhausts the land much less. If left for the seeds to mature they are very abundant and rich feed, but the stems are worthless, while the soil is more damaged.

Professor Phares says:

The matured stems are very hard, indigestible, and injurious, and the ripe seeds will founder more promptly than corn, and sometimes produce diabetes if moldy and too freely used. If cut at the right stage the whole plant is safe and very valuable forage. On good soil, if the ground be moist, it will be ready for mowing in sixty days from seeding, and produce from 2 to 4 tons of hay per acre. It is folly to sow it on poor land.

(Plate 19.)

Setaria glauca and Setaria viridis.

These two kinds, called pigeon grass, are very common in cultivated fields, especially among stubble after the cutting of grain. They are as nutritious as Hungarian grass but not so productive. (Plate 20.)

PENNISETUM.

The flowers in this genus are arranged in close spikes much like those of Setaria, but the bristles at the base of the spikelets fall off with the spikelets, instead of remaining attached to the rhachis.

Pennisetum spicatum (*Penicillaria spicata*) (Pearl Millet; Cat-tail Millet; Egyptian Millet).

This is supposed to be a native of Africa, but has been known from time immemorial in cultivation in India, Arabia, and Egypt.

It is an annual grass of luxuriant growth, frequently reaching 6 or 8 feet in height, with long, broad leaves, and a stout, solid culm terminated with a thick, erect spike, 6 to 10 inches long, and three-quarters of an inch in thickness, having a resemblance to the heads or spikes of the common cat-tail (*Typha latifolia*). The stalks are freely productive of suckers which furnish a large amount of succulent, sweet leaves.

Professor Phares states:

It has been grown to some extent for twenty-five years in many parts of the Southern States, more largely since 1865.

No crop will pay better or yield more forage than this on very rich, highly fertilized land. On such land it has been cut on an average every forty-five days, from the time of planting till frost, with a reported product of 60 or 100 tons of green forage, or from 16 to 20 tons of dry hay. When it grows luxuriantly it is impossible to cure it for hay on the ground upon which it is grown; so that it would be impracticable to make hay of a large field of it sown solid. Hence it must be sown in small patches or in beds, with spaces between upon which to spread it when cut. This difficulty would occur only on rich and highly manured land. Any one can have the crop as light as he chooses by sowing on poorly prepared or on exhausted land.

It is best adapted for cultivation in the South, where it will ripen seeds, but in a favorable season it may produce a large amount of forage in the Northern States.

TRIPSACUM.

Tripsacum dactyloides (Gama Grass).

A tall, stout, perennial grass, growing sparingly at the North, more common southward and in the Western States. The flowers are in spikes, generally from one to three at the top of the culm or from side shoots. The spikes are 2 to 4 inches long, the male flowers by themselves on the upper part, and the female flowers on the lower part. The lower flowers mature seeds in short joints, which break apart at maturity. Professor Phares says it was formerly found widely diffused through the Southern States, from the sea-shore to the mountains. It is now seldom seen, having been destroyed by cattle.

Mr Howard, of South Carolina, says of it:

This is a native of the South, from the mountains to the coast. The seed stem runs up to the height of 5 to 7 feet. The seeds break off from the stem as if from a joint, a single seed at a time. The leaves resemble those of corn. When cut before the seed stems shoot up they make a coarse but nutritious hay. It may be cut three or four times during the season. The quantity of forage which can be made from it is enormous. Both cattle and horses are fond of the hay. The roots are almost as large and strong as cane roots. It would require a team of four to six oxen to plow it up. It can, however, be easily killed by close grazing, and the mass of dead roots would certainly enrich the land. As the seeds of this grass vegetate with uncertainty, it is usually propagated by setting out slips of the roots about 2 feet apart each way. On rich land the tussocks will soon meet. In the absence of the finer hay grasses this will be found an abundant and excellent substitute. The hay made from it is very like corn fodder, is quite equal to it in value, and may be saved at a tithe of the expense.

(Plate 21.)

EUCHLAENA.

Euchlæna luxurians (Teosinte).

This grass is allied to and somewhat resembles Indian corn. Like it, it has the male flowers in a tassel at the top of the stalk, and the fertile ones arranged in slender spikes mostly concealed from view by the loose husk or sheath in which they are contained. These husks come from nearly every joint.

Prof. Asa Gray, in the American Agriculturist for August, 1880, speaking of this plant, writes:

The director of the botanic garden and Government plantations at Adelaide, Southern Australia, reports favorably of this strong-growing, corn-like forage plant, the *Euchlæna luxurians;* that the prevailing dryness did not injure the plants, which preserved their healthy green, while the blades of the other grasses suffered materially. The habit of throwing out young shoots is remarkable, sixty or eighty rising to a height of 5 to 6 feet. Further north, at Palmerston (nearer the equator), in the course of five or six months the plant reached the height of 10 to 14 feet, and the stems on one plant numbered fifty-six. The plants, after mowing down, grew again several feet in a few days. The cattle delight in it in a fresh state, also when dry. Undoubtedly there is not a more prolific forage plant known; but, as it is essentially tropic in its habits, this luxuriant growth is found in tropical or subtropical climates. The chief drawback to its culture with us will be that the ripening of the seed crop will be problematical, as early frosts will kill the plant. To make the teosinte a most useful plant in Texas, and along our whole Southern border, the one thing needful is to develop early flowering varieties so as to get seed before frost. And this could be done, without doubt, if some one in Texas or Florida would set about it. What it has taken ages to do in the case of Indian corn, in an unconscious way, might be mainly done in a human life-time by rightly directed care and vigorous selection. Who is the man who is going to make millions of blades of grass grow where none of any account ever grew before?

Seeds of this semi-tropical forage plant were distributed by the Department in the spring of 1886 and again in 1887. The plant considerably resembles Indian corn, but is more slender, gives off suckers more abundantly, and produces its seeds, a few together, in small tufts of husks instead of in ears. Each seed is inclosed by the peculiar hardened outer glumes, which would probably make it more difficult to digest than corn. The plant has not yet been extensively tried, owing to the difficulty of obtaining seed, which it was necessary to import, and which was therefore expensive and liable to be of poor quality. Experience has shown, however, that it will ripen in southern Florida, and in a few other favorable localities in the United States. Professor Phares, of Mississippi, believes, from instances that have come under his notice, that the seed may be successfully grown in some locations in the southern portion of that State, and over a considerable part of south-eastern Louisiana, and that in all parts of the Gulf States, even where it does not mature, it is destined to become a most valuable forage plant. It is probable that by selection and continued trial it may be made to ripen where it now does not.

J. C. Neal, Archer, northern Florida:

Often tried, and with much fertilizer makes a tremendous growth, giving a large amount of good forage, easily dried, and available. The seeds I received from the Department of Agriculture last year were deficient in vitality, and but few grew, but they showed that with good seed and care the teosinte would be a valuable forage plant. It will not ripen seed. I have tried to ripen it for ten years and failed.

J. G. Knapp, Limona, southern Florida:

Great difficulty has been experienced in obtaining live seed of this most valuable fodder plant, seed obtained from seedsmen, having been imported from Honduras, being too old to germinate. But during the past season a neighbor of mine has succeeded in obtaining a few seeds which grew, and his plants have matured their seeds, all of which will be planted the present year. Seed has also been matured at Fort Meade, in Polk County. Thus the question can be considered as settled, so far as this locality is concerned, that teosinte will mature its seed, and the country is placed in possession of the best soiling and fodder plant known to the agriculturists of the world. It endures heat, drought, and rains as well as sorghum and better than corn, and may be cured for hay.

Dr. Charles Mohr, Mobile, Ala.:

This tropical grass does not ripen its seeds in this latitude; it scarcely unfolds its blossoms before the advent of the first frost. It is very tender, being easily affected by frost or drought. During a cold spring it is difficult to secure a good stand, and it is only after warm weather has fairly set in that it begins to make a rapid growth, affording three cuttings and over of rich fodder on well manured ground in a season of genial showers. It is too succulent to be easily cured for hay. On that account and from the difficulty in securing a good stand and from the necessity of procuring each season a supply of seed from abroad, this grass has not found the favor with the cultivators of this section with which it is held in the subtropical zone.

J. S. Newman, Director Experiment Station, Auburn, Ala.:

Teosinte was cultivated on our experiment grounds last season with very satisfactory results. It tillers like cat-tail millet, but makes a much more luxuriant growth. It responds promptly and vigorously under the knife, and may be repeatedly cut during spring and summer. It does not, however, withstand drought as well as millo maize or kaffir corn, and it died completely during our seventy-five days of drought last fall. I have a few seeds which were matured on the grounds of Mr. George W. Benson, in the open air, at Marietta, Ga. He ripened seeds two years ago on a few plants which were forced in early spring and transplanted to the open ground. Last year this seed was planted in the open ground, and produced the plants which matured the seed which I have. He seems thus to have succeeded in acclimating the plant, which is therefore likely to prove a valuable acquisition.

(Plate 22.)

ZIZANIA.

Zizania aquatica (Wild Rice; Indian Rice).

Its ordinary growth is from 5 to 10 feet high, with a thick, spongy stem, and abundant long and broad leaves. The panicle is pyramidal in shape, 1 to 2 feet long, and widely branching below. The upper branches are rather appressed and contain the fertile flowers, while the lower branches contain only staminate ones. The spikelets are one-flowered, each with one pair of external husks or scales, which are by some botanists called glumes, and by others called palets. These husks or glumes in the fertile flower are nearly or quite an inch long, with an awn or beard as long, or twice as long. The grain inclosed between them is half an inch long, slender and cylindrical. The glumes of the staminate flowers are about half an inch long and without awns, each flower containing six stamens. These flowers fall off soon after they expand. The fertile flowers also drop very readily as soon as the grain is ripened.

This is botanically related to the common commercial rice (*Oryza sativa*) but is very different in general appearance. It is widely diffused over North America, and is found in Eastern Siberia and Japan. It grows on the muddy banks of rivers and lakes, both near the sea and far inland, sometimes in water 10 feet or more deep, forming patches or meadows covering many acres or extending for miles.

The grass abounds in the small lakes of Minnesota and the Northwest, and is there gathered by the Indians for food. The husk is removed by scorching with fire. It is a very palatable and nutritious grain. Some attempts have been made to cultivate the grass, but the readiness of the seed to drop must interfere with a successful result.

Near the sea-coast multitudes of reed-birds resort to the marshes, where it grows, and fatten upon the grain. The culms are sweet and nutritious, and cattle are said to be very fond of the grass. It is not adapted to culture on any ordinary farming land, as it will live only in the presence of water. (Plate 23.)

LEERSIA.

The flowers grow in spreading panicles. The spikelets are sessile, on short, one-sided branches or spikes. The spikelets are one-flowered, possessing but two scales, which may be called glumes or palets, which are strongly compressed, without awns, bristly ciliate on the keels, the lower one broader and inclosing the seed. Stamens one to six; stigmas two; grain flattened.

A genus of rough-leaved grasses growing for the most part in marshy or moist ground throughout nearly all parts of the United States. There are about five species, two of which are confined to the Southern States;

3594 GR——3

the others, at least two of them, are very common, though rarely occurring in great quantity. They are sometimes cut for hay. They can not be recommended for culture, but may be utilized wherever they grow spontaneously.

Leersia oryzoides (White Grass; Cut Grass; False Rice).

This is a handsome grass, the culms decumbent. It is commonly called rice grass, from its strong resemblance to common rice. The leaves are pale green, frequently a foot or more long, prominently veined below, very rough on the margins and on the sheaths. The panicle is about 1 foot long, diffusely branched, the branches mostly in twos, and an inch or two distant. The spikelets are very flat, about 2 lines long, nearly sessile, and borne mostly towards the ends of the long branches. The leaves are so rough on the margins as readily to cut the hand if roughly drawn through it.

Leersia Virginica (Small-flowered White Grass).

In this species the panicle is much smaller and narrower, and the branches appressed. The spikelets are smaller, the glumes narrower and smoother than in the first. (Plate 24.)

Leersia hexandra.

This species occurs in wet ground on the Atlantic and Gulf coast. It also occurs in other tropical and semi-tropical countries. It might be utilized in this country, if it becomes necessary, as it now is in some other countries. In Manilla, one of the Philippine Islands (as we learn from the United States consul at that place), this species is cultivated as food for horses and cattle. It is treated like rice, being transplanted to wet and previously plowed meadows. The local name there is *zacate*.

HILARIA.

Hilaria Jamesii (Gietta Grass).

This is one of the characteristic grasses of the arid districts of Texas, New Mexico, and Arizona, where it is sometimes called black grama. It is found sparingly also in Colorado and Utah. There are several other species growing in the same region, in some places quite abundantly. They are relished by cattle, and are considered as next in value to grama grass. (Plate 25.)

ANDROPOGON.

This genus is quite largely developed in the United States. They are perennial grasses, mostly tall, and with rough, wiry stems. Some of them occur in nearly all parts of the country from New England to Florida and west to Arizona. They are most abundant, however, in the Southern States, where they have been employed for permanent pastures. When they occur in quantity they can be utilized, but to be of value they should be kept from sending up their strong stems, as these are universally rejected by cattle and horses. Most of the species are not to be recommended for cultivation, but some have been praised in the South as furnishing, with proper management, permanent and reliable pastures.

Andropogon Virginicus and **Andropogon scoparius** (Broom Sedge).

Andropogon Virginicus and *A. scoparius* are the ones commonly employed in this way.

Dr. Charles Mohr, of Mobile, says that *Andropogon scoparius* grows extensively in old fields, and in the dry, sandy soil of the pine woods.

Much despised as it is as a troublesome weed, it has its good qualities, which entitle it to a more charitable consideration. In the dry pine woods it contributes, while green and tender, a large share to the sustenance of the stock.

It is common on the Western prairies, growing in dense tufts, and is known under the names of wire grass and bunch grass. It is, in most places, a constituent of prairie hay, and it makes good fodder if cut early. (Plates 26 and 27.)

Andropogon macrourus.

Andropogon macrourus, or heavy-topped broom grass, is frequent near the coast, from New Jersey to Florida, and thence west to Texas, and even to southern California. It has a stout culm, 3 to 4 feet high, with large, leafy clusters of flowers near the top. (Plate 28.)

Andropogon furcatus.

This is the tallest of our species. It grows erect to the height of 5 or 6 feet, in rocky or hilly ground; or at the West it is abundant on the native prairies, where it is frequently called blue stem. The leaves are long and frequently somewhat hairy on the sheaths and margins. The spikes are in small clusters of from three to six, terminating the stalk, and also with several clusters from the side branches. The spikes are usually 2 to 3 inches long, rather rigid, and contain ten to twenty flowers each. At each joint there is one sessile, perfect flower, and one stalked one, which is staminate only; otherwise it is nearly like the fertile one. The outer glumes are about four lines long, the upper one tipped with a short, stiff awn.

This species, as above stated, is abundant on the prairies of the West, where it is one of the principal hay grasses of the country, and is extensively cut and cured for winter use. (Plate 29.)

Andropogon Hallii.

This species much resembles the preceding, but the culms are stouter, the leaves thicker and more succulent, the flower spikes are larger, and the flowers generally more hairy. It prevails in very sandy soil, and is most frequent in western Kansas and in Colorado, Nebraska, and northward along the Missouri River. The leaves and stems are commonly of a light, bluish-green color. It will probably be well adapted to light, sandy soils.

CHRYSOPOGON.

Chrysopogon nutans (*Sorghum nutans*) (Wild Oats).

The stalks are 4 to 6 feet high, smooth, hollow, straight, and having at the top a narrow panicle, 6 to 12 inches long, of handsome straw-colored or brownish flowers, which is gracefully drooping at the top. The spikelets are at the ends of the slender branches of the loose panicle, generally of a yellowish color. At the base of each of the spikelets are two (one on each side) short, feathery pedicels; the flowers which they are supposed to have been made to support have entirely disappeared. The outer glumes are about three lines long, both alike, lanceolate, obtusish, coriaceous, five to seven-nerved, the lower one sparsely hairy, and with hairs at the base and on the stalk below.

This is a tall, perennial grass, having a wide range over all the country east of the Rocky Mountains. It grows rather sparsely and forms a thin bed of grass.

It is a nutritious grass, but should be cut early, as at full maturity the stems are coarse and are rejected by cattle. (Plate 30.)

SORGHUM.

In this genus the spikelets are much as in *Chrysopogon* and *Andropogon*, differing chiefly in habit and in the glumes of the fertile spikelets becoming hardened after flowering.

There are several species.

Sorghum halepense (Johnson Grass; Mean's Grass).

This grass is a native of Northern Africa and the country about the Mediterranean Sea.

It was introduced into cultivation in this country more than fifty years ago, and has recently attracted renewed attention, especially in the Southern States. The name Johnson grass, which is the one now most generally adopted in this country, originated from William Johnson, of Alabama, who introduced the grass into that State from South Carolina about the year 1840. It had previously been known as Mean's grass, and that name is still occasionally used. It has also been largely grown under the name of Guinea grass, but that name should be restricted to *Panicum maximum*, described in another part of this bulletin. It has been called Egyptian grass, Green Valley grass, Cuba grass, Alabama Guinea grass, Australian millet, and Morocco millet. In California it is best known as evergreen millet or Arabian evergreen millet. There seems to be good evidence that some of these names have been used at times in order to sell the seed as a new kind at an unreasonably high price. Johnson grass seeds abundantly, and the seed may be obtained of nearly all seedsmen under that name.

This grass is best adapted to warm climates, and has proved most valuable on warm, dry soils in the Southern States. It has been tested quite generally throughout the country, and is often recommended for cultivation even in the North, but there its growth is much smaller than at the South, and in severe winters it is killed outright. It is occasionally more or less winter-killed as far south as the northern portion of Texas and Alabama. Its chief value is for hay, in regions where other grasses fail on account of drought. If cut early the hay is of good quality, and several cuttings may be made in the season; but if the cutting is delayed until the stalks are well grown the hay is so coarse and hard that stock do not eat it readily. The seed may be sown at any time when the soil is warm and not too dry. Failures often occur from sowing the seed too early. If there is danger that the soil should dry out before the seed can germinate, soaking the seed may be resorted to with good results. Thick seeding gives a heavier yield and a better quality of hay. From 1 to 2 bushels are usually sown per acre, according to the cleanness of the seed. In case of failure to get a good stand the crop may be allowed to go to seed the first year, after which the vacant places will be found to be self-seeded. On small patches in such cases the ground is sometimes plowed up and the underground stems scattered along in the furrows over the vacant spots. In most localities it is generally considered desirable to plow the land set in Johnson grass about every third year; otherwise the root stocks become matted near the surface, and the crop is more affected by drought. Plowing causes it to grow more thickly and vigorously. If desired, a large portion of the root stocks may be removed at the time of plowing without injuring the stand. The greatest objection to Johnson grass is the difficulty of eradicating it. Care should be taken not to introduce it into fields intended for cultivation. It spreads rapidly, both by the root and by seed, and is apt to enter fields where it is not wanted. On stock-farms this feature is not so objectionable as elsewhere. The grass is not well adapted to pasture, and close pasturing is one of the means of getting rid of it. Its succulent, subterranean stems are usually well liked by hogs after they have become accustomed to them, and by keeping hogs closely confined upon it, it may be eradicated. Another method of eradication which is recommended is to plow in the fall, so as to expose it to the action of frost. In the South, where the grass is most largely grown, this is only partially successful.

There has been much discussion in the Southern papers respecting this grass, some considering it a great blessing, others a curse, the fact being that it is a blessing where a permanent grass is desired, and a great pest in land desired for general cultivation. It is probably too tender for the Northern States, but needs further trial. (Plate 31.)

Sorghum vulgare.

This name as at present applied includes several varieties quite different in appearance, as the variety *saccharatum*, or ordinary sugar sorghum, millo maize, Kaffir corn. dourra, and broom corn. The broom corn variety we need not discuss here. Some of the other varieties have been cultivated in various sections, and deserve especial attention in certain localities.

This plant has been widely discussed within the last few years in the agricultural press, and is valued by many who have grown it as a fodder plant in the South. There is considerable difference of opinion, however, as to its relative value as compared with the other sorghums and Indian corn. The following from among the replies received are given as additional evidence in regard to it:

J. S. Newman, Director Experiment Station, Agricultural and Mechanical College, Auburn, Ala.:

The popularity of this plant is waning, it having no special advantages over common corn, cat-tail millet, or common sorghum.

As evidence that millo maize has undergone acclimation, I will add that plants grown from seed freshly imported from South America do not mature seeds here.

Dr. Charles Mohr, Mobile, Ala.:

In the last three seasons this has been grown successfully in this vicinity by several parties. It ripens its seed before the advent of frost, which kills the plants to the roots. It does very well in the light soils of the coast plain, and perhaps everywhere in the pine region where there is a clay foundation. The growth of this grass during the early part of the season is much retarded by the chilly nights and spells of continued cold weather. It is only after the advent of settled warm weather that it enters upon its period of more vigorous growth.

Four cuttings may be taken during the season. Plants intended for seed are left undisturbed, and grow to a height of 18 or 20 feet, ripening in October. Great trouble in securing the seed is caused by the ravages of numerous birds.

The fodder obtained from the repeated cuttings, on account of its succulence, is difficult to cure, and in damp weather almost impossible. To cure dry fodder for winter use the plants are, after the second cutting, left to grow until towards the end of the season, when, having attained a height of 12 to 15 feet, and before opening their flowers, the stalks are cut and placed on end in small shocks. After being sufficiently dried they are placed upright under an airy shed or barn, protected from the damp. In this way sufficient ventilation is secured to prevent heating and molding, and to keep the fodder sweet and palatable. The fodder is said to be preferred by all kinds of live stock to any other fodder or hay. As to its nutritious value as compared with corn fodder opinions differ. The seeds are planted in spring in beds, which can be covered over during cool nights, and from these are transferred, when 8 to 10 inches in height, to the field, and thereafter treated in the same manner as corn.

Florida Farm and Fruit-Growers:

Red Millo Maize.—It stands drought and does not blow down easily, but it does not make as rank a growth as yellow millo maize. The seed is smaller than any of the other sorghums, and makes a first-class chicken-feed.

PHALARIS.

Phalaris arundinacea (Reed Canary Grass).

A perennial grass, with strong, creeping rhizomes, growing from 2 to 5 feet high, usually in low or wet ground. It ranges from New England and New York westward to Oregon, and northward to Canada, also in the mountainous parts of Pennsylvania and Virginia. It is common also in the north of Europe. The culm is stout, smooth, and leafy; the leaves are mostly from 6 to 10 inches long and about half an inch wide, the upper ones shorter.

The well-known ribbon grass of the garden is a variety of this grass, and will, it is said, easily revert to the normal type. In mountainous regions it may be worth trial for meadows. (Plate 32.)

Phalaris intermedia (Southern Reed Canary Grass; Gilbert's Relief Grass; Stewart's Canary Grass; California Timothy Grass).

This species resembles the foreign Canary grass (*Phalaris Canariensis*) which produces the seed commonly sold as food for Canary birds. It is, however, a taller and more robust species, growing 2 or 3 feet high, with a stout, erect culm and broad, linear leaves, which are from 4 to 10 inches long. The spike is oblong and compact, 1 or 2 inches long. There is a variety called var. *angusta*, in which the spike is 3 or 4 inches long. The spikelets are much like those of the preceding species (*Phalaris arundinacea*), having one perfect flower and two abortive ones. The outer glumes are lanceolate and nearly alike and have a narrow wing extending down the keel. The glumes of the fertile flower are nearly like those of *Phalaris arundinacea* already described.

This species grows in South Carolina and the Gulf States, extending to Texas, then stretching across to the Pacific coast and occurring through California and Oregon. It has frequently been sent to the Department from the Southern States as a valuable winter grass.

Mr. Thomas W. Beaty, of Conway, S. C., writes as follows:

The grass I send you was planted last September, and the specimens were cut on the 9th of March, following. You will notice that it is heading out and is just now in a right condition for mowing. It is wholly a winter grass, dying down in the latter part of April and first of May; and it seems to me should be a great thing for the South if properly introduced and cultivated, or rather the ground properly prepared and the seed sown at the right time. It would afford the best of green pasturage for sheep and cattle all winter. It is what we call Gilbert's relief grass.

Many years ago Dr. Lincecum, of Texas, experimented with this grass and recommended it very highly. In California it is called California Timothy, and is said to have little or no agricultural value. It is an annual or biennial. Professor Phares says:

The variety *angusta* is much larger and more valuable. It grows 2 to 3 feet high and in swamps 5 feet, with many leaves 4 to 10 inches long, the spike somewhat resembling the head of Timothy; stock like it well, especially as hay. Mr. D. Stewart, of Louisiana, having tested other grasses, prefers this for quantity and quality for winter and spring grazing, and for soiling for milk cows. There is much testimony from many parts of the South of the same import, and this grass is doubtless worthy of extended, careful testing.

(Plate 33.)

ANTHOXANTHUM.

Panicle somewhat spike-like. Spikelets apparently three-flowered, but only the terminal one perfect; the lower pair of glumes are equal, the lower one much smaller than the upper one; above these and below the perfect flower are two short, thin, two-lobed pubescent glumes, sometimes called abortive flowers, each one with an awn between the lobes; the upper or perfect flower is smaller, consisting of one broad, thin, three-nerved glume, and one (commonly considered the palet) narrow, one-nerved, hyaline glume. No true palet. Stamens two; styles two, distinct.

A. odoratum (Sweet Vernal Grass).

A perennial grass, native of Europe, much employed as a part of mixed lawn grasses, and frequently naturalized in meadows. It grows thinly on the ground, with slender culms, seldom more than 1 foot to 18 inches in height, and scanty in foliage. The panicle is 2 to 3 inches long, narrow, close, but expands considerably during flowering time. It is very fragrant and gives a pleasant odor to hay. (Plate 34.)

ALOPECURUS.

Alopecurus geniculatus (Water Foxtail).

This species and its variety *aristulatus*, which is the more abundant form, is native to this country. It commonly grows on the muddy banks of streams and lakes, and sometimes is found in wet meadows and ditches. It seldom reaches more than a foot in height; the stem is usually bent at the lower joint, and the sheaths of the leaves are more or less swollen, especially the upper one. It is of no value for cultivation, being useful only for the amount of grass it may contribute to the wild forage of the place in which it grows.

Alopecurus pratensis (Meadow Foxtail).

This is a perennial grass, a native of Europe, but it has been introduced into this country and is frequently found in meadows of the Eastern States. It has considerable resemblance to Timothy, but will be readily distinguished by an examination. It ordinarily grows but 2 feet high, but frequently in good soil reaches 3 feet or more. The culms are erect, with four or five leaves at pretty uniform distances. The sheaths are long and rather loose, particularly the upper one. The blade of the leaf is 3 or 4 inches long, about one-quarter of an inch wide at the base, and tapering gradually to a point. The panicle terminates the stalk, and is a cylindrical spike 2 or 3 inches long, dense, soft, and with the awns of the flowers conspicuously projecting. The spikelets are single-flowered, between 2 and 3 lines long. The outer glumes are strongly compressed, boat-shaped, keeled, nearly equal, sometimes slightly united together at the base, and have a line of soft, short hairs on the keels. These glumes closely inclose the flower, which is of nearly the same length, and consists of a flowering glume, but without any true palet. This flowering glume is folded upon itself and incloses the stamens and styles. It gives rise on its back, near the base, to a fine awn, which extends two or three lines beyond the glumes.

Mr. J. S. Gould says:

It flourishes in May, nearly four weeks in advance of Timothy, and is one of the earliest grasses to start in the spring. Pastures well covered with this grass will afford a full bite at least one week earlier than those which do not have it. It does not flourish in dry soils, but loves moist lands; no grass bears a hot sun better, and it is not injured by frequent mowings, on which account, as well as for its early verdure, it is valuable for lawns.

(Plate 35.)

Alopecurus occidentalis (Rocky Mountain Foxtail).

This species is indigenous in Montana and Idaho, and is very common along mountain streams, frequently covering acres of the mountain meadows. It is called in some localities mountain Timothy. It yields a large quantity of fine, bright hay, for which purpose it is often harvested and highly valued. It is of little value for grazing. Probably under cultivation it would become as useful as the European species.

ARISTIDA.

Spikelets one-flowered, in a spicate or an open branching panicle, generally on filiform pedicels; outer glumes unequal, often bristle-pointed; flowering glume narrow, rolled around the flower, terminating with a trifid awn, or apparently three-awned. Palet small and thin, inclosed in the flowering glume.

The grasses of this large genus are generally either worthless or of little agricultural value. The perennial species in some localities furnish a considerable amount of wild forage of an inferior character. They are very abundant in sandy and sterile soil, especially in the Rocky Mountain region.

Aristida purpurea.

Aristida purpurea prevails extensively on the Western plains, and it is said to form an important part of the early feed of the cattle. It grows in bunches, and is about 1 foot high. The panicle is somewhat spreading, and the flowers are purplish, with spreading, slender awns, 1 inch or more in length. (Plate 36.)

STIPA.

Spikelets one-flowered, terete, spicate, or paniculate. Outer glumes membranaceous, keeled; flowering glume narrow, coriaceous, rigid, involute, with a simple twisted awn from the apex; palet usually small and inclosed by the flowering glume. Stamens generally three. The flowering glume has a hardened, often sharp-pointed and bearded pedicel or stipe at its base.

This genus has its principal range in the region of the Rocky Mountains and the Great Plains. They are mostly coarse, rigid grasses, having little agricultural value. In common with many other kinds they are usually called bunch grass, sometimes beard grass, or feather grass. The more abundant species are *Stipa spartea*, *Stipa comata*, and *Stipa viridula*. These prevail from British America southward, on the plains, and in the mountains. The genus is particularly distinguished by the awn or beard of the flowering glume, and the sharp-pointed and barbed *stipe* or base of the glume. Complaint has been made among stockmen of great injury to sheep by the penetration of these sharp points into the wool, and even into the flesh. The awns or feathery appendages are in some species 4 to 6 inches long, and are subject to a spiral twisting when dry, which assists in forcing the seed into the ground for germination. *Stipa arenacea* is the only species prevailing in the Eastern and Southern States, and is of no agricultural importance. The long, feathery awns of *Stipa pennata* are beautiful and ornamental. (Plate 37, *Stipa viridula*.)

Stipa spartea.

Stipa spartea is called porcupine grass, arrow grass, and devil's knitting-needles, from the long, stiff, twisted awns inclosing the seed. The seeds ripen early and drop to the ground, and later in the season the grass may be easily recognized by the persistent, bleached culms and empty glumes of the spreading panicle. The long root-leaves continue green and vigorous throughout the summer; frequently being 2 feet long. Although somewhat coarse the grass makes a very good hay, and forms a considerable part of the wild prairie hay in Iowa, Nebraska, Minnesota, and southern Dakota. It is called buffalo grass in the Saskatchewan region. It should receive attention in Western experiments for a pasture grass. (Plate 38.)

ORYZOPSIS.

This genus differs from *Stipa* chiefly in having a shorter ovate or oblong flower, with the callus at the base shorter and broader, and in having usually a very short and deciduous awn to the flowering glume.

Oryzopsis cuspidata (Bunch Grass).

This grass has a wide distribution, not only in the Sierras of California, but northward to British America, and eastward through all the interior region of Utah, Nevada, New Mexico, Texas, Colorado, and Nebraska, to the Missouri River. It is a perennial, growing in dense tufts, whence its common name.

The culms are 1 to 2 feet high, with about three narrow, convolute leaves, the upper one having a long, inflated sheath which incloses the base of the panicle. The radical leaves are narrow, rigid, and as long as or longer than the culm. The panicle is about 6 inches long, very loose, spreading and flexuous. The branches are in pairs, slender, rather distant, and are subdivided mostly in pairs. The spikelets are at the ends of the capillary branches, each one-flowered. The outer glumes are 3 to 4 lines long, inflated and widened below, gradually drawn to a sharp-pointed apex, thin and colorless except the three or five green nerves, and slightly hairy. The glumes inclose an ovate flower, which is covered externally with a profusion of white, silky hairs, and tipped with a short awn, which falls off at maturity. This apparent flower is the flowering glume, of a hard, coriaceous texture, and incloses a similar hard, but not hairy, and smaller palet.

In Montana it is one of the most esteemed bunch grasses, and thrives on soil too sandy for other valuable species. Professor Brewer states that in southern California it is called *saccatoo* or *saccatoa*. (Plate 39.)

MILIUM.

Spikelets panicled ; outer glumes membranaceous, equal and convex, the flowering glume and its palet coriaceous, much as in *Panicum*, but the articulation with the rhachis is above the outer glumes. All the glumes are unawned, and there is no sterile pedicel.

Milium effusum.

A tall, perennial grass, 4 or 5 feet high, growing in damp woods in the northern portions of the United States and in Canada. It is also found in Northern Europe and in Russian Asia.

Hon. J. S. Gould, in the Report of the New York State Agricultural Society, says, respecting this grass:

Meadows and borders of streams and cold woods. It thrives when transplanted to open and exposed situations. It is one of the most beautiful of the grasses; the panicle is often a foot long, and the branches are so exceedingly delicate that the small, glossy spikelets seem to be suspended in the air. Birds are very fond of the seed. Mr. Colman says that he has raised 3 tons to the acre of as good, nutritious hay as could be grown from it, when sown in May. The plants multiply by the roots as well as by the seed, sending out horizontal shoots of considerable length, which root at the joint as they extend.

(Plate 40.)

MUHLENBERGIA.

Spikelets one-flowered, small, paniculate, articulated above the outer glumes; flowering glume with a very short, usually hairy callus.

Muhlenbergia diffusa (Nimble Will).

Professor Killebrew, of Tennessee, says:

It is hardly more than necessary to mention this grass, which forms in many sections the bulk of the pastures of the woods. It does not grow in fields, but in woods, where, after rains have set in, it carpets the earth with living green. Various opinions are entertained as to its nutritive qualities. Some farmers assert that their stock are fond of it, and that on sufficient range, cattle, horses and sheep will go into the winter sleek and fat from this vigorous grass. Others regard it as well nigh worthless.

(Plate 41.)

Muhlenbergia glomerata (Spiked Muhlenbergia).

This grass grows in wet, swampy grounds, chiefly in the Northern and Western portions of the United States. It is found in Colorado, Utah, Nevada, New Mexico, and Texas. It grows to the height of 2 or 3 feet, stiffly erect and generally unbranched.

The culm is hard, somewhat compressed, and very leafy. The panicle is narrow, 2 to 4 inches long, composed of numerous close clusters of flowers, becoming looser below, forming an interrupted glomerate spike. The spikelets are closely sessile in the clusters. The root-stock is hard and knotty, and furnished with numerous short, firm shoots or stolons.

In the Eastern States it is utilized as one of the native products of wet meadows in the making of what is called wild hay. Specimens have been sent from Colorado and Kansas, and recommended as an excellent grass for forage. (Plate 42.)

Muhlenbergia Mexicana.

A perennial grass of decumbent habit, 2 or 3 feet high, much branched, from scaly, creeping root stocks. It is frequently found in moist woods and low meadows or prairies. It probably would not endure upland culture, but in its native situations it fills an important place among natural grasses. (Plate 43.)

Muhlenbergia sylvatica.

This species has much the appearance and habit of *Muhlenbergia Mexicana*. The panicle is looser, the spikelets not so densely clustered, and the flowering glume bears an awn two or three times as long as itself. It is found in dry, open, or rocky woods and fence corners. In agricultural value it corresponds to that species. (Plate 44.)

PHLEUM.

Phleum pratense (Herd's Grass [of New England and New York]).

This is one of the commonest and best-known grasses. For a hay crop it is extensively cultivated, especially in the Northern and Western States. The height of the grass depends on the soil and cultivation. In poor ground it may be reduced to 1 foot, while in good soil and with good culture it readily attains 3 feet, and occasionally has been found twice that height. It is a perennial grass, with fibrous roots.

The base of the culm is sometimes thickened and inclined to be bulbous. The culm is erect and firm, with four or five leaves, which are erect, and usually 4 to 6 inches long. The flower spike is cylindrical and very densely flowered, and varies from 2 to 6 inches in length. The spikelets are sessile, single-flowered, and cylindrical or oblong in outline. The outer glumes are rather wedge-form, with a mucronate point or short bristle. The main nerve on the back is fringed with a few short hairs.

This grass, as known in cultivation, is supposed to have been introduced from Europe, but the earliest account that we have of its culture is that given by Jared Elliot, who says it was found by a Mr. Timothy Herd in a swamp in New Hampshire, and that he began its cultivation. As it was found to be a valuable grass, its cultivation soon spread, and it was known as Herd's grass.

It was not introduced into cultivation in England until some fifty years later. I consider it very probable that the specimens found by Mr. Herd were of native growth, for it is believed to be native in the White Mountains, in the Rocky Mountains, in Alaska, and in Labrador. It is also a native of Europe. But, however the question of its nativity may be settled, the thanks of this country are due to Mr. Herd for the introduction into agriculture of one of the most valuable of grasses. It is now a favorite meadow grass over a large part of the country, and its hay is a staple, and more sought after in the markets than any other kind.

Timothy thrives best on moist, loamy soil of medium tenacity, and is not suited to light, sandy, or gravelly soils. Under favorable circumstances and with good treatment it yields very large crops, often four tons to the acre. One writer states that he has known whole fields in Missouri grow to the height of 5 or 6 feet, the soil, a pulverized clay, being particularly suited to this grass. He also states that he has known fields of this grass to be highly productive for thirteen years in succession. Farmers should not lose sight of the fact that the roots do not extend widely, and that much of its vitality depends on the thickened bulb-like base of the stem; therefore there is danger that, if mowed too late in the season so that the bulbs and roots are left unprotected from the weather, they may suffer from the action of frost, being sometimes lifted out of the ground from this cause.

A well-informed farmer, writing in the Prairie Farmer, says that Timothy is an exhaustive crop, the roots not penetrating deeply enough to obtain nourishment from the subsoil. Feeding off with stock lays the crown of the plants bare, which, being of a bulbous nature, are easily injured by exposure. When, however, the aftermath is very abundant, Timothy meadows may be pastured sparingly in the fall to reduce the heavy growth of rowan that sometimes accumulates so as to interfere with the mower; but in no case should sheep be allowed upon it, as they are very apt to nip off the crown of the plant and thus destroy it. In order to keep up the productiveness of a Timothy meadow, a good top-dressing of s able manure should be applied and evenly spread in the fall. This will protect the roots and cause a much thicker and stronger growth. Timothy is often sown with clover in different proportions, and under some circumstances this is a judicious practice. But the more general practice is to have the Timothy meadows free from other plants, and to sow about 12 pounds of seed to the acre.

When this grass is grown for the crop of seed, it should be allowed to stand until the heads are ripe; 30 bushels to the acre have been produced. Of course the hay left after thrashing out the seed is coarse and of inferior value. The clean seed weighs about 45 pounds to the bushel. (Plate 45.)

SPOROBOLUS.

Spikelets one, rarely two-flowered, in a contracted or open panicle; outer glumes unequal, the lower one shorter, often acute, unawned, one to three-nerved, membranaceous; flowering glume mostly longer, unawned; palet about equaling the flowering glume and of the same texture, prominently two-nerved. Seed mostly loose in a hyaline or rarely coriaceous pericarp.

Sporobolus cryptandrus.

This species has an extensive range. It is common in sandy fields in the Northern and Southern States, as well as over all the dry plains west of the Mississippi, extending from British America to Mexico, furnishing a considerable share of the wild pasturage of that region. (Plate 46.)

Sporobolus Indicus.

This grass is a native of India, but has spread over most tropical and warm climates. It occurs more or less abundantly in all the Southern States, and is called smut grass, from the fact that after flowering the heads become affected with a black smut. It grows in tufts or loose

46

patches, from 1½ to 3 feet high, with an abundance of long, flat, fine-pointed leaves at the base, and a narrow terminal panicle, frequently a foot in length, composed of short, erect, sessile branches, which are very closely flowered.

Professor Phares says:

It grows abundantly and luxuriantly on many uncultivated fields and commons, and furnishes grazing from April till frost. It thrives under much grazing and many mowings, and grows promptly after each if the soil is moist enough. Cattle and horses are fond of it, if it is frequently cut or grazed down, but if allowed to remain untouched long they will not eat it unless very hungry, as it becomes tough and unpalatable and probably difficult to digest.

(Plate 47.)

AGROSTIS.

Spikelets one-flowered, in a contracted or open panicle; outer glumes nearly equal or the lower rather longer, and longer than the flowering glume, one-nerved, acute, unawned; flowering glume shorter and wider, hyaline, three to five-nerved, awnless or sometimes awned on the back; palet shorter than the flowering glume, frequently reduced to a small scale or entirely wanting; stamens usually three; grain free.

Agrostis vulgaris (Redtop, Finetop, Herd's Grass [of Pennsylvania], Bent Grass, etc.)

A perennial grass, growing 2 or 3 feet high from creeping root-stocks, which interlace so as to make a very firm sod; the culms are upright, or sometimes decumbent at the base, smooth, round, rather slender and clothed with four or five leaves, which are flat, narrow, and roughish, from 3 to 6 inches long, with smooth sheaths, and generally truncate ligules. It is extensively cultivated.

Agrostis alba, the florin grass of Ireland, and *Agrostis stolonifera* are usually considered synonymous, and are distinguished from *Agrostis vulgaris* by having a closer, more verticillated panicle and with longer and more acute ligules.

Mr. J. G. Gould says of *Agrostis vulgaris:*

This is a favorite grass in wet, swampy meadows, where its interlacing, thick roots consolidate the sward, making a firm matting which prevents the feet of cattle from poaching. It is generally considered a valuable grass in this country, though by no means the best one. Cattle eat hay made from it with a relish, especially when mixed with other grasses. As a pasture grass it is much valued by dairymen, and in their opinion the butter would suffer much by its removal.

Professor Phares, of Louisiana, says, respecting this grass:

It grows well on hill-tops and sides, in ditches, gullies and marshes, but delights in moist bottom-land. It is not injured by overflow, though somewhat prolonged. It furnishes considerable grazing during warm spells in winter, and in spring and summer an abundant supply of nutriment. Cut before maturing seed, it makes hay and a large quantity. It seems to grow taller in the Southern States than it does farther north, and to make more and better hay and grazing.

Mr. Flint says:

It is a good permanent grass, standing our climate as well as any other, and consequently well suited to our pastures, in which it should be fed close; for if allowed to grow up to seed the cattle refuse it; and this seems to show that it is not so much relished by stock as some of the other pasture grasses.

(Plate 48.)

47

Agrostis canina.

A grass usually of low size, 6 to 12 inches high, with slender culms, and a light, flexible, expanded panicle, and with a perplexing variety of forms. There are several varieties growing in mountainous regions throughout the United States, and in Europe. It forms a close sod, and affords considerable pasturage in those regions. It is probably one of the grasses called Rhode Island bent grass.

Agrostis exarata.

This is chiefly a northwestern species, being found in Wisconsin and westward to the Rocky Mountains, also from British America and California to Alaska. It is very variable in appearance, and presents several varieties. It is generally more slender in growth than the common redtop. The panicle is usually longer, narrower, and looser. In all the forms the palet is wanting or is very minute. The form chiefly growing on the Pacific slope from California to Alaska is often more robust than the *Agrostis vulgaris*, growing 2 to 3 feet high, with a stout, firm culm, clothed with three or four broadish leaves, 4 to 6 inches long. The panicle is 4 to 6 inches long, pale green, rather loose, but with erect branches.

It deserves trial for cultivation, at least on the Pacific side of the continent. (Plate 49.)

CINNA.

Spikelets one-flowered, much flattened, in an open, spreading panicle; outer glumes lanceolate, acute, strongly keeled, hispid on the keel, the upper somewhat longer than the lower; flowering glume manifestly stalked above the outer glumes, about the same length as the outer ones, three-nerved, short-awned on the back near the apex; palet nearly as long as its glume, only one-nerved (probably by the consolidation of two, Bentham); stamen one. A sterile pedicel sometimes present.

Cinna arundinacea (Wood Reed Grass).

A perennial grass, with erect simple culms from 3 to 6 feet high, and a creeping rhizoma; growing in swamps and moist, shaded woods in northern or mountainous districts. The leaves are broadly linear lanceolate, about 1 foot long, 4 to 6 lines wide, and with a conspicuous elongated ligule. The panicle is from 6 to 12 inches long, rather loose and open in the flower, afterwards more close.

This leafy-stemmed grass furnishes a large quantity of fodder, but experiments are wanting to determine its availability under cultivation. (Plate 50.)

Cinna pendula.

This species is more slender, with a looser drooping panicle and more capillary branches, and with thinner glumes. It occurs in the same situations as the preceding, and is more common in the Rocky Mountains and Oregon.

AMMOPHILA.

Spikelets one-flowered, in a contracted, spike-like or open, diffuse panicle, with or without a bristle-like rudiment opposite the palet; outer glumes large, nearly equal, rigid, thick, lanceolate, acute, keeled, five-nerved; flowering glume similar in texture, about equal in length, sometimes mucronate at the apex; palet as long as its glume, of similar texture, two-keeled, sulcate between the keels; hairs at the base of the flower usually scanty and short.

Ammophila arundinacea (Beach Grass; Sand Grass).

This is *Calamagrostis arenaria* of the older books. The entire plant is of a whitish, or pale-green color. It grows on sandy beaches of the Atlantic, at least as far south as North Carolina, and on the shores of the Great Lakes, but has not, so far, been recorded from the Pacific coast. It also grows on the sea-coast of the British Isles and of Europe. It forms tufts of greater or less extent, "its long, creeping roots extending sometimes to the extent of 40 feet, and bearing tubers the size of a pea, interlaced with death-like tenacity of grasp, and form a net-work beneath the sand which resists the most vehement assault of the ocean waves." The culms are from 2 to 3 feet high, rigid and solid; the leaves long, involute, smooth, stiff, and slender-pointed; the panicle is dense, 6 to 10 inches long, close and spike-like; the spikelets are about half an inch long, compressed, crowded on very short branchlets.

This grass has no agricultural value, but from time immemorial its utility in binding together the loose sands of the beach, and restraining the inroads of the ocean, has been recognized and provided for in some places by law. Mr. Flint, in his work on grasses, says that the town and harbor of Provincetown, once called Cape Cod, where the Pilgrims first landed, one of the largest and most important harbors of the United States, sufficient in depth for ships of largest size, and in extent to anchor three thousand vessels at once, owe their preservation to this grass. The usual way of propagating the grass is by transplanting the roots. It is pulled up by hand and placed in a hole about a foot deep and the sand pressed around it by the foot. There are undoubtedly many places on the sea-coast where this grass would be of inestimable value in restraining the encroachment of the ocean. It would also be useful in forming a dense turf for the protection of dikes and banks subject to water-washing.

CALAMAGROSTIS.

This genus is characterized by having one-flowered spikelets, with the addition at the base of the flowering glume of a small hairy appendage or pedicel, which is considered to be the rudiment of a second flower. In addition to this the flower is also generally surrounded at the base with a ring of soft hairs, and the flowering glume usually bears an awn on its back, which is generally bent and twisted.

In this genus there are two sections, viz: 1st, Deyeuxia, in which there is a small hairy pedicel in front of the palet of the single perfect flower; the glumes thin and membranaceous. In this section are most of our North American species. 2d, Calamovilfa, in which the glumes and palet are thicker and more compressed, and the sterile pedicel or rudiment is wanting.

Calamagrostis (Deyeuxia) Canadensis (Blue-joint; Small Reed Grass).

A stout, erect, tall perennial grass, growing chiefly in wet, boggy ground or in low, moist meadows. Its favorite situation is in cool, elevated regions. It prevails in all the northern portions of the United States, in the Rocky Mountains, and in British America. In those districts it is one of the best and most productive of the indigenous grasses.

It varies much in luxuriance of foliage and size of panicle, according to the location.

The culms are from 3 to 5 feet high, stout and hollow, hence in some places it is called the small reed grass. The leaves are 1 foot or more long, flat, from a quarter to nearly half an inch wide, and roughish; the stem and sheaths smooth.

The panicle is oblong in outline, open, and somewhat spreading, especially during flowering; it is from 4 to 6 or even 8 inches in length, and 2 or 3 inches in diameter, of a purplish color; the branches are mostly in fives at intervals of an inch or less. These branches vary in length from 1 to 3 inches, the long ones flowering only toward the extremity. The spikelets are short-stalked, the outer glumes about one and one-half lines long, lanceolate and acute; the silky white hairs at the base of the flowering glume are about as long as the glume; those on the sterile pedicel also nearly as long. The flowering glume is thin and delicate, about as long as the outer glumes, and somewhat finely toothed at the apex, three to five-nerved, and bearing on the back, below the middle, a delicate awn, reaching about to the point of the glume, and not much stouter than the hairs. The proper palet is thin, oblong, and about two-thirds the length of its glume.

Mr. J. S. Gould says:

It constitutes about one-third of the natural grasses on the Beaver Dam Meadows of the Adirondacks. It is certain that cattle relish it very much, both in its green state and when made into hay, and it is equally certain that the farmers who have it on their farms believe it to be one of the best grasses of their meadows.

Professor Crozier, who spent some time in northwestern Iowa and adjacent parts of Minnesota and Dakota, in studying the native grasses, says:

This is considered by some to produce the best hay for cattle of all the native grasses. It is very leafy, and stands remarkably thick on the ground. The seed ripens early in July, but the leaves remain green until winter. It is probably hardly equal to some of the upland grasses in quality, but it gives a larger yield, and is undoubtedly worthy of cultivation. It is usually found upon the margins of ponds; it will thrive, however, on land that is only slightly moist, and often occurs along the banks of roadside ditches. On rather low land which has been broken and allowed to go back it frequently comes in, and after a few years occupies the land to the exclusion of all other vegetation.

(Plate 51.)

Calamagrostis (Deyeuxia) sylvatica (Bunch Grass).

A coarse perennial grass, growing in large tufts, usually in sandy ground in the Rocky Mountains at various altitudes, also in California, Oregon, and British America. It furnishes an abundant coarse forage in the regions where it is found. The culms are from 1 to 2 feet high, erect, rigid, and leafy; the radical leaves are frequently as long as the culm, two or three lines wide, sometimes flat, but generally involute and rigid. The culm leaves are from 3 to 6 or 8 inches long, rigid and rough. The panicle is narrow and spike-like, 3 to 5 inches long, erect, rather dense, sometimes interrupted below, and varying from pale green to purple. (Plate 52.)

Calamagrostis longifolia.

This grass grows on the sandy plains of the interior from British America to Arizona, and on the borders of the Great Lakes. It has strong, running rootstocks, like

3594 GR——4

the preceding, but is much taller, the culms being 3 to 6 feet high, stout and reed-like; the leaves long, rigid, and becoming involute, with a long, thread-like point. The panicle is quite variable, from 4 to 16 inches long, at first rather close, but becoming open and spreading, the branches in the smaller forms being 2 or 3 inches long, and in the larger ones often 10 or 12 inches and widely spreading. It is abundant on the plains of western Nebraska, Kansas, and Colorado, and furnishes a resource in winter for food for the cattle of the ranches.

(Plate 53.)

HOLCUS.

Spikelets two-flowered, crowded in an open panicle, the lower flower perfect, the upper one male only, and with a minute, hairy rhachilla or rudiment at its base. Outer glumes nearly equal, compressed, membranaceous, large (fully inclosing the two flowers), flowering glumes half shorter, the lowest awnless, the upper with a short dorsal awn.

Holcus lanatus (Velvet Grass; Velvet Mesquite; Soft Grass, etc.).

Introduced from Europe and naturalized in many parts of the United States. It makes a striking and beautiful appearance, but stock are not very fond of it, either green or cured. It is a perennial, but not very strongly rooted, and does not spread from the root as do most perennial grasses. It seeds abundantly, and is generally propagated by seed, though sometimes by dividing the plants. It prefers low land, but does very well even on sandy upland, and its chief value is in being able to grow on land too poor for other grasses.

The seed has been in market many years, but it has come into cultivation very slowly, and it is not generally held in very high esteem as an agricultural grass, either in this country or in Europe. Some speak well of it, however, and it has frequently been sent to the Department from the South, with strong recommendations for its productiveness.

C. Menelas, Savannah, Ga.:

Known almost all over the South as yielding more than orchard grass, but for some reason only grown where nature has planted it.

Mrs. J. W. Bryan, Dillon, northwestern Georgia:

My meadows and ditches are full of it, though it is not sown here. It is very valuable for pasture, and gives a very early and heavy yield of hay.

L. S. Nicholson, Crumly, northeastern Alabama:

This grass has been grown on a farm I own for about ten years. It does best on rich, moist land, but grows fairly well on poor, dry, sandy land, where other and, I must say, better grasses fail.

It grows from 2 to 3 feet high, and makes apparently sufficient hay, but very light and chaffy and of inferior quality. It appears to be hardy and will withstand drought well. The grass is right pretty when growing, and nice for pasturing, but we have other grasses so much better that can generally be grown on land that this would occupy that I shall vote against it for all purposes.

Clarke Lewis, Cliftonville, Miss.:

It grows on poor, sandy land to a height of 3 to 4 feet; stands drought well, but can be killed by a slight overflow. It is valuable as a soiling plant, but makes inferior hay. It is an annual, and if intended for a permanent meadow must be cut only once, and then allowed to reseed itself.

H. W. L. Lewis, secretary Louisiana State Grange, Tangipahoa Parish, La. (P. O., Osyka, Miss.):

It is hardy and cultivated in small lots, doing best on rich, sandy loam, yielding 2 to 3 tons per acre. I have experimented more than any one else in my section with forage plants, especially winter grains and grasses. Have used rye and barley for winter feed, but have given them up in favor of the *Holcus lanatus;* have had this in cultivation for thirty years. It is a perennial, but owing to its shallow roots it dies out during our long, dry summer and fall from 50 to 75 per cent. One lot kept the third year had less than 10 per cent. of the grass alive. Hence I have for twenty years or more used it as an annual, sowing it with turnips, collards, or by itself. A good way is to sow the seed broadcast and cover lightly in a late crop of turnips after the last cultivation. After the turnip crop is removed the first warm days in January or February will start the grass into rapid growth. It is cut frequently through the spring for green feed, and after oats are ready to cut is allowed to mature seed.

Prof. D. L. Phares, in his "Farmer's Book of Grasses," says:

In the Eastern States this grass is called Salem grass and white Timothy; in the South, velvet lawn grass, and velvet mesquite grass; in England, woolly soft grass and Yorkshire white. It has been sent to me for name more frequently than any other grass. Having found its way to Texas, people going there from other States have sent back seeds to their friends, calling it Texas velvet mesquite grass, supposing it a native of that State. So far as has come to my knowledge nine-tenths of all so-called mesquite grass planted in the Southern States is this European velvet grass. It grows much larger in some of the Southern States than in the Eastern States or in England, and seems to have greatly improved by acclimation.

Velvet grass may be readily propagated by sowing the seed or dividing the roots. It luxuriates in moist, peaty lands, but will grow on poor, sandy, or clay hill lands and produce remunerative crops where few others will make anything. The reason that cattle do not prefer it is not because of a deficiency in nutrition, but because of its combination. It is deficient simply in saline and bitter extractive matter which cattle relish in grasses.

It is by no means the best of our grasses, but best on some lands. Other grasses are more profitable to me. It should be sown from August to October, 14 pounds equal to 2 bushels per acre. Northward it is perennial, in the South it is not strictly so.

(Plate 54.)

TRISETUM.

Spikelets two to three, rarely five-flowered, in a dense or open panicle, the rachis usually hairy and produced into a bristle at the base of the upper flower; outer glumes unequal, acute, keeled, membranaceous, with scarious margins; flowering glumes of similar texture, keeled, acute, the apex two-toothed, the teeth sometimes prolonged into bristle-like points, the middle nerve furnished with an awn attached above the middle, which is usually twisted at the base and bent in the middle; palet hyaline, narrow, two-nerved, two-toothed.

Trisetum palustre.

A slender grass, usually about 2 feet high, growing in low meadows or moist ground throughout the eastern part of the United States. The culms are smooth, with long internodes and few linear leaves, 2 to 4 inches long; the panicle is oblong, 3 to 4 inches long, loose and gracefully drooping, the branches two to five together, rather capillary, 1 to

1½ inches long and loosely flowered; the spikelets are two-flowered; the outer glumes are about two lines long, the lower one one-nerved, the upper rather obovate and three-nerved; the lower flower is commonly awnless or only tipped with a short awn; the second flower is rather shorter and with a slender, spreading awn longer than the flower.

This is a nutritious grass, but is seldom found in sufficient quantity to be of much value. (Plate 55.)

Trisetum subspicatum.

The culms are erect and firm, smooth or downy. The panicle is spike-like, dense, and cylindrical or elongated, and more or less interrupted, generally of a purplish color. The spikelets are two or three-flowered. The flowers are a little longer than the outer glumes, slightly scabrous, the flowering glumes acutely two-toothed at the apex, and bearing a stout awn which is longer than its glume.

A perennial grass of the mountainous region of Europe and North America; undoubtedly furnishes a considerable portion of mountain pasturage. It is found sparingly in New England, on the shores of Lake Superior, in the Rocky Mountains of Colorado, Utah, California, Oregon, and northward to the Arctic circle. It varies in height according to the latitude at which it grows, being sometimes reduced to 3 or 4 inches, at other times running up to 2 feet high. (Plate 56.)

AVENA.

Avena fatua (Wild Oats).

This species is very common in California. It is generally thought to have been introduced from Europe, where it is native, but it has become diffused over many other countries, including Australia and South America. It is thought by some to be the original of the cultivated oat, *Avena sativa*, that the common will degenerate into the wild oat, and that by careful cultivation and selection of seed the wild oat can be changed into the common cultivated oat. But on this question there is a conflict of opinions, and the alleged facts are not sufficiently established. The wild oat differs from the cultivated one chiefly in having more flowers in the spikelets, in the long, brown hairs which cover the flowering glumes, in the constant presence of the long, twisted awn, and in the smaller size and lighter weight of the grain. It is a great injury to any grain-field in which it may be introduced; but for the purposes of fodder, of which it makes a good quality, it has been much employed in California. (Plate 57.)

ARRHENATHERUM.

Arrhenatherum avenaceum (Evergreen Grass; Meadow Oat Grass; Tall Oat Grass).

Culms 2 to 4 feet high, erect, rather stout, with four or five leaves each; the sheaths smooth, the leaves somewhat rough on the upper surface, 6 to 10 inches long, and about 3 lines wide, gradually pointed. The panicle is loose, rather contracted, from 6 to 10 inches long, and rather drooping; the branches very unequal, mostly in fives, the longer ones 1 to 3 inches, and subdivided from about the middle; the smaller branches very short, all rather full-flowered. The spikelets are mostly on short pedicels. The structure of the flowers is similar to that of common oats, but different in several particulars. The spikelet consists of two flowers, the lower of which is staminate only, the upper one both staminate and pistillate; the outer glumes are thin and transparent, the upper ones about 4 lines long and three-nerved, the lower one nearly 3 lines long and one-nerved. The flowering glume is about 4 lines long, green, strongly seven-nerved, lanceolate, acute, hairy at base, roughish, and in the lower flower gives rise on the back below the middle to a long, twisted, and bent awn; in the upper flower the glume is merely bristle-pointed near the apex. The palet is thin and transparent, linear and two-toothed.

This grass is much valued on the continent of Europe. The herbage is very productive and its growth rapid. When growing with other grasses, cattle and sheep eat it very well, but do not like to be confined to it exclusively. It is a perennial grass of strong, vigorous growth, introduced from Europe and sparingly cultivated.

Professor Phares, of Mississippi, says:

It is widely naturalized and well adapted to a great variety of soils. On sandy or gravelly soils it succeeds admirably, growing 2 to 3 feet high. On rich, dry upland it grows from 5 to 7 feet high. It has an abundance of perennial, long, fibrous roots penetrating deeply in the soil, being therefore less affected by drought or cold, and enabled to yield a large quantity of foliage, winter and summer. These advantages render it one of the very best grasses for the South, both for grazing, being evergreen, and for hay, admitting of being cut twice a year. It is probably the best winter grass that can be obtained. It will make twice as much hay as Timothy. To make good hay it must be cut as soon as it blooms, and after it is cut must not be wet by dew or rain, which damages it greatly in quality and appearance. For green soiling it may be cut four or five times, with favorable seasons.* In from six to ten days after blooming the seeds begin to ripen and fall, the upper ones first. It is therefore a little troublesome to save the seed. As soon as those at the top of the panicle ripen sufficiently to begin to drop, the panicle should be cut off and dried, when the seeds will all thrash out readily and be matured.

After the seeds are ripe and taken off, the long, abundant leaves and stems are still green, and being mowed make good hay. It may be sown in March or April and mowed the same season; but for heavier yield it is better to sow in September or October. Not less than 2 bushels (14 pounds) per acre should be sown. The average annual nutriment yielded by this grass in the Southern belt is probably twice as great as in Pennslyvania and other Northern States.

A. P. Rowe, Fredericksburgh, Va.:

Tall oat grass has been seeded here and does well. It comes in with orchard grass for hay, and the two might be seeded together with the best results.

T. W. Wood & Sons, Richmond, Va.:

It is cultivated very generally for pasture and hay; and is the best grass we know for thin soils. It is hardy, stands drought moderately well, is easily subdued, and lasts five or six years.

Dr. W. J. Beal, Agricultural College, Michigan:

It is cultivated in a few places in the State, proving perfectly hardy, and doing best on deep, porous soils where it stands drought very well, yielding perhaps 3 tons per acre. It makes good pasture, and lasts a long time.

J. J. Dotson, Cedarton, Tex.

It is very fair for early spring pastures, and to cut for green feed when it first heads, in March, but it is not liked as hay. It is too light and the seeds fall off too easily. I have never known it cultivated. Thrives only on low bottom-land.

(Plate 58.)

CYNODON.

Cynodon Dactylon (Bermuda Grass).

A low, creeping perennial grass, with abundant short leaves at the base, sending up slender, nearly leafless, flower stalks or culms, which have three to five slender, diverging spikes at the summit. The spikelets are sessile in two rows on one side of the slender spikes; they each have one flower, with a short-pediceled, naked rudiment of a second flower; the outer glumes nearly equal, keeled; the flowering glume boat-shaped, broader, and prominently keeled; the palet narrow, and two-keeled.

This is undoubtedly, on the whole, the most valuable grass in the South. It is a native of Southern Europe, and of all tropical countries. It is a common pasture-grass in the West Indies and the Sandwich Islands, and has long been known in the United States, but the difficulty of eradicating it when once established has retarded its introduction into cultivation. Its value, however, is becoming more appreciated now that more attention is being given to grass and relatively less to cotton, and better methods and implements of cultivation are being employed. Still, it seems probable, from the reports received, that at the present time a majority of farmers would prefer not to have it on their farms. It seeds very sparingly in the United States, and as the imported seed is not always to be had, and is expensive, and often of poor quality, those who have desired to cultivate it on a large scale have seldom been able to do so. It is generally used as a lawn grass, and to hold levees or railroad embankments, and for small pastures. In some localities, however, it has spread over a considerable extent of territory. Its natural extension into new territory has been slow, owing to the partial or entire absence of seed, but it spreads rapidly by its rooting stems when introduced. It is usually propagated artificially by means of the sets or rooting stems. These are sometimes chopped up with a cutting-knife, sown broadcast, and plowed under not very deeply; sometimes they are dropped a foot or two apart in shallow furrows, and covered by a plow; sometimes pieces of the sod are planted two feet apart each way. By any of these means a continuous sod is obtained in a few months if the soil is good and well prepared.

The chief value of Bermuda grass is for summer pasture. It grows best in the hottest weather, and ordinary droughts affect it but little. The tops are easily killed by frosts, but the roots are quite hardy throughout the Southern States. It is grown to some extent as far north as Virginia, but in that latitude it possesses little advantage over other grasses. In Tennessee, according to Professor Killebrew, its chief value is for pasture, there being other grasses there of more value for hay. Farther South, however, it is highly prized for hay. To make the largest quantity and best quality it should be mowed several times during the season. The yield varies greatly according to soil, being generally reported at from a ton and a half to two tons per acre. Much larger yields have been reported, however, in specially favorable localities where several cuttings were made.

Bermuda grass is more easily eradicted from sandy land than from clay, and on such land may be more safely introduced into a rotation. To kill it out it should be rooted up or plowed very shallowly some time in December and cultivated or harrowed occasionally during the winter. If severe freezes occur most of it will be killed by spring; or it may be turned under deeply in spring and the land cultivated in some hoed crop or one which will heavily shade the ground.

M. M. Martin, Comanche, Comanche County, central Texas:

Bermuda grass grows on any kind of soil in Texas, but will not stand the tramping of stock on loose, sandy soil. It is hard to beat for a grazing grass, though long droughts cause it to dry up. It is not very early to start in the spring.

Whitfield Moore, Woodland, Red River County, northeastern Texas:

Bermuda stands droughts well, is a good fertilizer, grows well from fifteen to twenty years from one planting, then only needs plowing to renew it. It is tolerably easily subdued by shallow turning in early winter, so that it will freeze. It yields heavy crops of hay and can be mowed three times a year. It is the finest grass I have ever seen for summer grazing, and when inclosed from stock during the summer it is fine winter grazing. It will stop washing, and cause low, wet land to fill up and become dry.

E. W. Jones, Buena Vista, Miss.:

Bermuda has been a great terror to planters until recently. If plowed shallow late in the fall, and allowed to freeze during winter, there is no trouble to cultivate a crop the next season. The ground becomes perfectly mellow, and though the grass is not dead, it does but little injury to the crop.

E. Taylor, Pope's Ferry, Ga.:

Nothing kills it except severe freezing. It is the best of all grasses, and thrives on any soil, but best on clay. It furnishes good pasture from May until the middle of November. For winter grazing bur clover is taking its place. The yield of hay is about 2 tons per acre. It will reclaim the poorest lands, and is not very difficult to subdue. It ripens seeds in this State sparingly.

J. B. Wade, Edgewood, DeKalb County, northern Georgia:

This is about the most northern limit at which Bermuda grass grows in this State. It is beginning to be highly appreciated both for grazing and for hay. It stands drought well, keeping green from May until November. It makes good hay, and can be cut two or three times a year, producing on an average 2½ tons of hay per acre. While this is the most northern limit of Bermuda grass, it is also the most southern limit of blue grass. The two growing together on the same land produce a most perfect pasture, as the blue grass is green nearly all the fall, winter, and spring months, while during the heat of summer, which prevents the growth of the blue grass, the Bermuda flourishes. The two together in good, strong soil make a perfect pasture, good all the year around.

Prof. S. M. Tracy, now Director of the Mississippi Agricultural Experiment Station, formerly of the Agricultural College, Columbia, Mo.:

It has been in cultivation near St. Louis, in one locality only, for many years. It barely survives the winter and would doubtless be destroyed by pasturing. I have noticed it very carefully about New Orleans, where it is by far the most valuable permanent pasture grass, and is thoroughly naturalized, if not a native. It is almost the only grass grown there for winter pasture or for lawns. It stands drought well, and grows anywhere except on very wet ground. It can be subdued by one year of thorough cultivation.

Prof. J. B. Killebrew, in "The Grasses of Tennessee," says:

Occasionally the traveler meets with patches of Bermuda grass in the cotton fields of the South, where it is carefully avoided by the planter, any disturbance giving new start to its vigorous roots. Some ditch around it, others inclose it and let shrubbery do the work of destruction. It forms a sward so tough that it is almost impossible for a plow to pass through it. It will throw its runners over a rock 6 feet across and hide it from view, or it will run down the sides of the deepest gully and stop its washing. It does not, however, endure shade, and in order to obtain a good stand the weeds must be mown from it the first year. It would be a good grass to mix with blue grass, as when it disappears in winter the blue grass and white clover would spring up to keep the ground in a constant state of verdure. This experiment has been tried with eminent success. It grows luxuriantly on the top of Lookout Monutain, having been set there many years ago. This mountain is 2,200 feet high, and has, of course, excessively cold winters.

(Plate 59.)

SPARTINA.

A genus of coarse, perennial grasses, growing mainly in marshy grounds, from strong, scaly root stocks. The flowers are produced in one-sided spikes of the panicle. The spikelets are closely sessile, and mostly crowded on the triangular axis. They are one-flowered, and much flattened laterally. The empty glumes are unequal strongly compressed and keeled, acute, the keel mostly hispid, the upper one longer than the lower; flowering glume compressed and keeled, awnless; palet about equaling its glume.

Spartina cynosuroides (Cord Grass).

A coarse and stout grass, growing from 3 to 5 feet high, with leaves 2 to 3 feet long. The top of the culm for about 1 foot is occupied by from five to ten spikes, which are from 1½ to 3 inches long, and the spikelets are very closely imbricated. The lower glume is linear-lanceolate, the upper one lanceolate with a long, stiff point.

This species has a wide range, from near the coast to the base of the Rocky Mountains. In the Western States it is very common, often forming a large part of the grass of the sloughs and wet marshes of that region. It is frequently cut for hay, but is of inferior quality unless cut very early.

In the bottom-lands of the Mississippi, where it is abundant, it has been manufactured into paper. (Plate 60.)

Spartina juncea (Salt Grass; Marsh Grass).

A much smaller species than the preceding, usually 1 to 2 feet high, from a creeping, scaly root stock, the culms rigid and the leaves nearly round and rush-like. There are from two to five spikes, which are 1½ to 2 inches long and on short peduncles.

This grass forms a large portion of the salt-marshes near the coast. It makes an inferior hay, called salt hay, which is worth about half as much per ton as Timothy and redtop. It is much employed as a packing material by hardware and crockery dealers. (Plate 61.)

BOUTELOUA.

(GRAMA GRASS.)

Spikes single or numerous in a racemose, commonly one-sided panicle; spikelets commonly densely crowded in two rows on one side of the rhachis, each consisting of one perfect flower and a stalked pedicel bearing empty glumes and one to three stiff awns; outer glumes unequal, acute, keeled, membranaceous; flowering glume broader, usually thicker, with three to five lobes, teeth, or awns at the apex; palet narrow, hyaline, entire or two-toothed, infolded by its glume.

Bouteloua oligostachya (Grama Grass; Mesquite Grass).

This is the commonest species on the great plains. It is frequently called buffalo grass, although that name strictly belongs to another plant (*Buchloë dactyloides*). On the arid plains of the West it is the principal grass and is the main reliance for the vast herds of cattle which are raised there. It grows chiefly in small, roundish patches closely pressed to the ground, the foliage being in a dense, cushion-like mass. The leaves are short and crowded at the base of the short stems. The flowering stalks seldom rise over a foot in height, and bear near the top one or two flower-spikes, each about an inch long, and from one-eighth to one-quarter of an inch wide, standing out at right angles like a small flag floating in the breeze. Where much grazing prevails, however, these flowering stalks are eaten down so much that only the mats of leaves are observable. In bottom-lands and low, moist ground it grows more closely, and under favorable circumstances forms a pretty close sod, but even then it is not adapted for mowing, although it is sometimes cut, making a very light crop. Under the most favorable circumstances the product of this grass is small compared with the cultivated grasses. It is undoubtedly highly nutritious. Stock of all kinds are fond of it and eat it in preference to any grass growing with it. It dries and cures on the ground so as to retain its nutritive properties in the winter. No attempt is made by stockmen to feed cattle in the winter; they are expected to "rustle around," as the phrase is, and find their living; and in ordinary winters, as the fall of snow is light, they are enabled to subsist and make a pretty good appearance in the spring; but in severe winters there are losses of cattle, sometimes very heavy ones, from want of feed. (Plate 62.)

Bouteloua racemosa (Mesquite Grass; Tall Grama Grass).

This species ranges from Mexico to British America and east of the Mississippi River, in Wisconsin and Illinois. It is easily distinguished from the others by its taller growth and by the long, slender raceme of twenty to fifty or more slender spikes. These are usually about half an inch long and reflexed. There are from six to ten spikelets on each spike. Although eaten by cattle, especially when made into hay, it is not so much relished as some other kinds.

There are about a dozen other species of this genus occurring more or less extensively in the Southwest, chiefly in New Mexico and Arizona, all of which are nutritious grasses, but seldom occurring in sufficient quantity to be particularly important. (Plate 63.)

ELEUSINE.

Spikes two to five or more, finger-like, at the summit of the culm, sometimes a few scattering ones lower down; spikelets sessile and crowded along one side of the rhachis; two to six (or more)-flowered, the uppermost flowers imperfect or rudimentary; outer glumes membranaceous, shorter than the spikelet; flowering glumes usually obtuse; palet folded, two-keeled.

Eleusine Indica (Yard Grass; Crow-foot; Crab Grass; Wire Grass).

The culms are from 1 to 3 feet high, usually coarse and thick, and very .eaty, especially below. The leaves are long and rather wide. At the top of the culm there are two to five or more thickish densely-flowered spikes proceeding from a common point, with sometimes one or two scattering ones lower down on the clum. The spikelets are sessile and crowded along one side of the axis, each being from two to six-flowered, the upper flower imperfect or rudimentary; the outer glumes are membranaceous, shorter than the flowers, the flowering glumes usually obtuse; the palet folded and two-keeled.

An annual grass belonging to tropical countries, but now naturalized in most temperate climates. In the Southern States it is found in every door-yard and in all waste places.

Professor Phares, of Mississippi, says:

The clumps have many long leaves and stems rising 1 or 2 feet high, and many long, strong, deeply-penetrating, fibrous roots. It grows readily in door-yards, barnyards, and rich, cultivated grounds, and produces an immense quantity of seeds. It is a very nutritious grass, and good for grazing, soiling, and hay. The succulent lower part of the stems, covered with the sheaths of the leaves, render it difficult to cure well, for which several days are required. It may be cut two or three times, and yields a large quantity of hay.

(Plate 64.)

Eleusine Ægyptiaca (Crow-foot.)

Two species of grass in the Southern States have received the name of crow-foot, viz: *Eleusine Indica* and *Eleusine Ægyptiaca*, or, as it is sometimes called, *Dactyloctenium Ægyptiacum*. Dr. H. W. Ravenel, of Aiken, S. C., states that in the lower and middle portions of that State the name of goose grass is generally applied to the former, while the latter is universally called crow foot. *E. Indica*, he says, is confined to rich waste places and old yards and gardens, and is rarely or never seen in ordinary cultivated fields, and is never used for hay, because it is found only in tufts and sparsely, whilst *E. Ægyptiaca*, is as abundant as crab grass (*Panicum sanguinale*) in all cultivated fields, and it is commonly used for hay.

This is an important distinction, which ought to be generally known and noticed in our popular account of these grasses. (Plate 65.)

BUCHLOE.

Buchloë dactyloides (Buffalo Grass).

This grass is extensively spread over all the region known as the Great Plains. It is very low, the bulk of leaves seldom rising more than 3 or 4 inches above the ground, growing in extensive tufts, or patches, and spreading largely by means of stolons or off-shoots similar to those of the Bermuda grass, these stolons being sometimes 2 feet long, and with joints every 3 or 4 inches, frequently rooting and sending up flowering culms from the joints. The leaves of the radical tufts are 3 to 5 inches long, one or one-half line wide, smooth or edged with a few scattering hairs. The flowering culms are chiefly diœcious, but sometimes both male and female flowers are found on the same plant, but in separate parts. Next to the grama grass it is, perhaps, the most valuable plant in the support of the cattle of the plains. (Plate 66.)

TRIODIA.

Spikelets several to many-flowered in a strict spike-like or an open, spreading panicle, some of the upper flowers male or imperfect; outer glumes keeled, acute or acutish, awnless; flowering glumes imbricated, rounded on the back, at least below, hairy or smooth, three-nerved, either mucronate, three-toothed, or three-lobed at the apex, or obscurely erose, often hardened, and nerveless in fruit; palet broad, prominently two-keeled.

Triodia sealerioides (Tall Redtop).

This grass grows from 3 to 5 feet high. The culms are very smooth; the leaves are long and flat, the lower sheaths hairy or smoothish.

The panicle is large and loose, at first erect, but finally spreading widely. The branches are single or in twos or threes below, and frequently 6 inches long, divided and flower-bearing above the middle. The spikelets are on short pedicels, 3 to 4 lines long and five or six-flowered. The outer glumes are shorter than the flowers, unequal and pointed; the flowering glumes are hairy towards the base, having three strong nerves, which are extended into short teeth at the summit. It is a large and showy grass when fully matured, the panicles being large, spreading, and of a purplish color.

It grows in sandy fields, and on dry, sterile banks, from New York to South Carolina and westward. It is eaten by cattle when young, but the mature culms are rather harsh and wiry and not relished by them. It is, however, cut for hay where it naturally abounds.

The genus *Triodia* has its chief distribution in Texas and the adjacent region, where there are several species which seem to have some importance in the grass supply of these arid districts. Among these are *Triodia trinerviglumis, Triodia stricta, Triodia Texana*, and *Triodia acuminata.*

These deserve further investigation. (Plate 67.)

ARUNDO.

Tall grasses with an ample panicle, spikelets two to many-flowered, the flowers rather distant, silky-hairy at the base, and with a conspicuous silky-bearded rhachis, all perfect; outer glumes narrow, unequal, glabrous, lanceolate, keeled, acute; flowering glumes membranaceous, slender, awl-pointed; palets much shorter than the glumes, two-keeled, pubescent on the keels.

Arundo Donax (Giant Reed Grass).

This grass is often cultivated for its very ornamental plumes. It is a native of Southern Europe, but is well established on the borders of the Rio Grande River, where it is probably indigenous, and has been recommended for cultivation.

PHRAGMITES.

Only differing from *Arundo* in the lowest flower of the spikelets being staminate only and glabrous.

Phragmites communis (Reed Grass).

A tall, coarse, perennial grass, growing on the borders of ponds and streams, almost rivaling sorghum in luxuriance. It attains a height of 6 to 10 feet; the culms sometimes an inch in diameter, and leaves an inch or two in width. The panicle is from 9 to 15 inches long, loose, but not much spreading, of an oblong or lanceolate form, and slightly nodding. The branches are very numerous, irregularly whorled, 4 to 8 inches long, much subdivided, and profusely flowering. The largest panicles form very ornamental plumes, almost equal to those of *Arundo Donax*, so much cultivated for ornamental purposes. It sometimes attains the height of 15 feet. It is resorted to by cattle only when finer and more nutritious grasses fail. (Plate 68.)

KŒLERIA.

Kœleria cristata.

This grass has a very wide diffusion, both in this country and in Europe and Asia. It favors dry hills or sandy prairies, and on the Great Plains is one of the commonest species. It occurs throughout California and extends into Oregon. It varies much in appearance, according to the location in which it grows, these varieties being so striking that they have been considered different species; and perhaps two species ought to be admitted. It is perennial, with erect culms usually from 1 to 2 feet high, and a spike-like panicle varying from 3 to 6 inches in length, and more or less interrupted or lobed at the lower part. When grown in very arid places the culms may be only a foot high, the radical leaves short, and the panicle only 2 inches long. When grown in more favored situations the radical leaves are sometimes 18 inches long, the stem 3 feet, and the panicle 6 inches. The branches of the panicle are, in short, nearly sessile clusters, crowded above, looser and interrupted below. The spikelets are from two to four-flowered. On the prairies west of the Mississippi it is one of the commonest and most useful of the grasses. In Montana it is sometimes called June grass. It is an early grass, ripening about the first of July. (Plate 69.)

ERAGROSTIS.

Spikelets several; usually many-flowered, pedicellate or sessile, in a loose and spreading, or narrow and clustered panicle; the rhachis of the spikelets usually glabrous and articulate under the flowering glumes, but often tardily so, and sometimes inarticulate. Outer empty glumes unequal, and rather shorter than the flowering ones, keeled, one-nerved; flowering glumes obtuse or acute, unawned, three-nerved, the keel prominent, the lateral nerves sometimes very faint; palet shorter than the glume, with two prominent nerves or keels, often persisting after the glume and grain have fallen away.

Eragrostis major.

This is a foreign grass which has become extensively naturalized, not only in the older States, but in many places in the Western and Southwestern Territories. It is found in waste and cultivated grounds, and on roadsides, growing in thick tufts, which spread out over the ground by means of the geniculate and decumbent culms. The culms are from 1 to 2 feet long, the lower joints bent and giving rise to long branches. The sheaths are shorter than the internodes, the leaves from 3 to 6 inches long. The panicle is frequently 4 or 5 inches long, oblong or pyramidal, somewhat open, but full-flowered; the branches single or in pairs, branched and flowering nearly to its base. This grass is said to have a disagreeable odor when fresh. It produces an abundance of foliage, and is apparently an annual, reaching maturity late in the season. We are not aware that its agricultural value has been tested. (Plate 70.)

Eragrostis Abyssinica.

Eragrostis Abyssinica is a species which has been introduced from Abyssinia, and cultivated in Florida and some of the Southern and Southwestern States, and is said to be remarkably productive and valuable for hay. It is an annual grass, growing to the height of 2 to 3 feet. The native Abyssinian name of this grass is "*teff*," and from the seeds the Abyssinians make their bread. It may be cultivated with ease at a height of 6,000 or 7,000 feet above the sea-level, where maize can hardly thrive. It comes to maturity in four months, yields forty times its volume of seed, and, in the best variety, is said to make a white, delicious bread. The traveler Bruce mentions *teff* with approval, and there is some account of it in other books. The Royal Gardens of Kew obtained a quantity of seed, of which they sent a portion to the U. S. Department of Agriculture, and by the Department it has been distributed to the agricultural stations for trial. There are many other species, but none of much agricultural importance.

DISTICHLIS.

Distichlis maritima (Salt Grass; Akaline Grass).

It has strong, creeping root stocks covered with imbricated leaf-sheaths, sending up culms from 6 to 18 inches high, which are clothed nearly to the top with the numerous, sometimes crowded, two-ranked leaves. The leaves are generally rigid and involute, sharp-pointed, varying greatly in length on different specimens. The plants are diœcious, some being entirely male and some female. The panicle is generally short and spike-like, sometimes, especially in the males, rather loose, with longer, erect branches, and sometimes reduced to a few spikelets. The spikelets are from 4 to 6 inches long and five to ten-flowered, the flowers being usually much compressed. The outer glumes are smooth, narrow, and keeled; the flowering ones are broader, keeled, acute, rather rigid, and faintly many-nerved. The palets have an infolded margin, the keels prominent or narrowly winged. The pistillate spikelets are more condensed and more rigid than the staminate.

This is described in most botanical works as *Bryzopyrum spicatum*, but recently the name given by Rafinesque has been accepted and restored to it by Mr. Bentham. It is a perennial grass, growing in marshes near the sea-coast on both sides of the continent and also abundantly in alkaline soil throughout the arid districts of the Rocky Mountains.

Although this can not be considered a first-rate grass for agricultural purposes, it is freely cut with other marsh grasses, and on the alkaline plains of the Rocky Mountains it affords an inferior pasturage. (Plate 71.)

DACTYLIS.

Dactylis glomerata (Orchard Grass).

The culm and leaves roughish, the leaves broadly linear, light green, and five to six on the culm. The panicle is generally but 2 or 3 inches long, the upper part dense from the shortness of the branches; the lower branches are longer and spreading, but with the spikelets glomerated or closely tufted. The spikelets are usually three to four-flowered, one-sided, and on short, rough pedicels. The glumes are pointed and somewhat unequal, the upper one being smaller and thinner than the lower. The flowering glumes are ovate-lanceolate, roughish, and ending in a sharp point or short awn, and are rather longer than the outer glumes.

This is one of the most popular meadow grasses of Europe, and is well known to most farmers in the Northern and Eastern States. It is a perennial of strong, rank growth, about 3 feet high.

Professor Phares, of Mississippi, says:

Of all grasses this is one of the most widely diffused, growing in Africa, Asia, every country of Europe, and all our States.

It is more highly esteemed and commended than any other grass, by a large number of farmers in most countries, a most decided proof of its great value and wonderful adaption to many soils, climates, and treatments. Yet, strange to say, though growing in England for many centuries, it was not appreciated in that country till carried there from Virginia in 1764. But, as in the case of Timothy grass, soon after its introduction from America, it came into high favor among farmers, and still retains its hold on their estimation as a grazing and hay crop. It will grow well on any soil containing sufficient clay and not holding too much water. If the land be too tenacious, drainage will remedy the soil; if worn out, a top dressing of stable manure will give it a good send-off, and it will furnish several mowings the first year. It grows well between 29 degrees and 48 degrees latitude. It may be mowed from two to four times a year, according to latitude, season, and treatment, yielding from 1 to 3 tons of excellent hay per acre on poor to medium land. It is easily cured and handled. It is readily seeded and catches with certainty. It grows well in open lands and in forests of large trees, the underbrush being all cleared off. I know but one objection to it. Like tall oat grass it is disposed to grow in clumps and leave much of the ground uncovered. This may be obviated by thick seeding, using 2¼ or, better, 3 bushels of seed per acre. The gaps may be prevented by sowing with it a few pounds of redtop seed. But as the latter multiplies annually from seeds dropping, it would in a few years root out the orchard grass. In common with many others I prefer red clover with orchard grass. It fills the gaps and matures at the same time with the orchard grass; the mixture makes good pasture and good hay; but if mowed more than twice a year, or grazed too soon after the second mowing, the clover will rapidly fail. One peck of red clover seed and 6 pecks of orchard grass seed is good proportion per acre.

After being cut it has been found to grow 4 inches in less than three days. Sheep leave all other grasses if they can find this, and acre for acre it will sustain twice as many sheep or other stock as Timothy. Cut at the proper age it makes a much better hay than Timothy, and is greatly preferred by animals, being easier to masticate, digest, and assimilate; in fact more like green grass in flavor, tenderness, and solubility.

Mr. J. S. Gould, of New York, says:

The testimony that has been collected from all parts of the world for two centuries past establishes the place of this species among the very best of our forage grasses, and we have not a shadow of a doubt that the interests of our graziers and dairymen would be greatly promoted by its more extended cultivation. It is always found in the rich old pastures of England, where an acre of land can be relied on to fatten a bullock and four sheep. It is admirably adapted for growing in the shade, no grass being equal to it in this respect, except the rough-stalked meadow grass (*Poa trivialis*). It receives the name of orchard grass from this circumstance. We have seen it growing in great luxuriance in dense old New England orchards, where no other grass except *Poa trivialis* would grow at all. It affords a good bite earlier in the spring than any other grass except the meadow foxtail (*Alopecurus pratensis*). It affords a very great amount of aftermath, being exceeded in this respect by no other grass except Kentucky blue grass (*Poa pratensis*), and it continues to send out root-leaves until very late in the autumn. When sown with other grasses its tendency to form tussocks is very much diminished; indeed it is always unprofitable to sow it alone in meadows or pastures, as it stands too thin upon the ground to make a profitable use of the land, and the filling up of the interspaces with other varieties greatly improves the quality of the orchard grass by restraining its rankness and making it more delicate.

From Colman's Rural World:

Orchard grass makes good winter pasturage, equally as good as blue grass, and far better pasturage in seasons of drought than blue grass, as it is a deeper and larger-rooted plant and resists drought better. When once established it can be fed as closely as any other grass, and is no harder on land than any other. Indeed, land pastured in orchard grass will continue to improve in fertility. If half of each of our farms were well seeded to orchard grass it would be a great advantage to them.

From the Farmer's Home Journal:

This is one of the most valuable of all the grasses, and is better adapted to the South than any other with which we are acquainted. Its rapidity of growth and the luxuriance of its aftermath, its power of enduring drought and the cropping of cattle, commended it highly to the farmer, especially as a pasture grass, and it is rapidly growing in favor. It starts earlier in the spring, and continues growing later in the fall, and starts again more quickly after being cut, than any other grass, thus furnishing both the earliest and latest grazing. Orchard grass is less exhausting to the soil than Timothy. It will endure considerable shade. In a porous subsoil its fibrous roots extend to a great depth. It does well on any soil of even moderate fertility which is not too wet for grass, and will grow and thrive where no other grass will. It does best on a sandy loam with a porous subsoil, but will grow on a sand-bank if made rich enough. When sown alone, we would sow 2 bushels to the acre. From the nature of its growth thick seeding is necessary to secure the best results, and we think the farmer will never regret the extra first cost of sowing two bushels per acre.

When sown thickly and properly protected from grazing it forms a close and very durable turf. Nothing will hurt it except plowing. As to time of sowing, it may be sown in August, September, October, February, March, or April, alone, or on wheat, rye, or oats. Hay made from a mixture of this grass with clover is very nutritious, second only to best Timothy hay made, falling very little behind it, while in most lands in the South the yield will exceed that of Timothy.

Orchard grass is ready for grazing in the spring ten or twelve days sooner than any other that affords a full bite. When grazed down and the stock turned off, it will be ready for regrazing in less than half the time required for Kentucky blue grass.

(Plate 72.)

POA.

Spikelets somewhat compressed, usually two to five-flowered, in a narrow or loose and spreading panicle, the rhachis between the flowers glabrous or sometimes hairy, the flowers generally perfect, in a few species diœcious; outer glumes commonly shorter than the flowers, membranaceous, keeled, obtuse or acute, one to three-nerved, not awned; flowering glumes membranaceous, obtuse ar acute, five or rarely seven-nerved, the intermediate nerves frequently obscure, often scarious at the apex and margins, smooth or puboscent, often with a few loose or webby hairs at the base; palet about as long as the flowering glumes, prominently two-nerved or two-keeled.

Poa arachnifera (Texas Blue Grass).

This species was first described by Dr. John Torrey in the report of Captain Marcy's exploration of the Red River of Louisiana, as having been found on the headwaters of the Trinity, and named *Poa arachnifera* from the profuse webby hairs growing about the flowers, although it is found that this character is very variable, probably depending somewhat on the amount of shade or exposure to which the grass is subjected.

Several years ago Mr. Hogan, of Texas, sent specimens of the grass to this Department, and as it was shown to be a relative of the Kentucky blue grass, Mr. Hogan adopted for the common name Texas blue grass. We give some extracts from his letters relating to the grass:

I find it spreading rapidly over the country, and I claim for it all and more in Texas than is awarded to the *Poa pratensis*. It seems to be indigenous to all the prairie country between the Trinity River and the Brazos in our State. It blooms here about the last of March, and ripens its seeds by the 15th of April. Stock of all kinds and even poultry seem to prefer it to wheat, rye, or anything else grown in winter. It seems to have all the characteristics of *Poa pratensis*, only it is much larger, and therefore affords more grazing. I have known it to grow 10 inches in ten days during the winter. The coldest winters do not even nip it, and although it seems to die down during summer, it springs up as soon as the first rains fall in September, and grows all winter. I have known it in cultivation some five years, and have never been able to find a fault in it. It will be ready for pasture in three or four weeks after the first rains in the latter part of August or 1st of September. I have never cut it for hay. Why should a man want hay when he can have green grass to feed his stock on?

Mr. James E. Webb, of Greensborough, Hale County, Ala., writing to the Department December 26, 1888, says:

Recent experiments show that the Texas blue grass (*Poa arachnifera*) flourishes and grows here in west Alabama as finely as could be wished, and is likely ere long to furnish us what we so much need, a fine winter grass. With Texas blue grass, Melilotus and Bermuda grass, Alabama is a fine stock country.

Mr. S. C. Tally, of Ellis County, Texas, has sent specimens of this grass. He says it is abundant there, bears heavy pasturing, and makes a beautiful yard or lawn grass.

Similar favorable accounts have been received from others. It is likely to prove one of the most valuable grasses for the South and Southwest. By means of its strong stolons or offshoots it multiplies rapidly and makes a dense, permanent sod. It produces an abundance of radical leaves which often become 2 feet in length, and those of the culms are smooth and of good width, about 4 to 8 inches long and 2 lines wide. The culms are 2 to 3 feet high, each with two or three leaves, with long sheaths and blade, the upper leaf sometimes reaching nearly to the top of the panicle. The ligule is round and short, or lacerated when old. The panicle is from 3 to 8 inches in length, rather narrow, and with short, erect branches of equal length, in clusters of from three to five, the longest seldom 2 inches, most of them short, some nearly sessile and profusely flowering to the base. The spikelets usually contain about five flowers.

In many cases there is a remarkable development of long, silky hairs at the base of each flower, but sometimes these are quite absent. (Plate 73.)

Poa compressa (English Blue Grass; Wire Grass).

This species has sometimes been confounded with the Kentucky blue grass, from which it differs in its flattened, decumbent, wiry stems, its shorter leaves and shorter, narrower, and more scanty panicle. It is found in many old pastures, on dry banks, and in open woods. The culms are hard and much flattened, 1 foot to 18 inches long, more or less decumbent, and frequently bent at the lower joints. The leaves are scanty, smooth, short, and of a dark, bluish-green color. The panicle is short and contracted, 1 to 3 inches long. Very contradictory accounts have been given as to its agricultural value, some denouncing it as worthless and others entertaining a good opinion of it. It thrives well on clay or hard, trodden, and poor soils.

Hon. J. S. Gould says, respecting it:

It is certain that cows that feed upon it both in pasture and in hay give more milk and keep in better condition than when fed on any other grass. Horses fed on this hay will do as well as when fed on Timothy hay and oats combined.

These discrepant opinions may be due in part to having mistaken the *Poa pratensis* for this grass. It is probably a nutritious grass, but from its spare yield can hardly obtain much favor for a hay crop. (Plate 74.)

3594 GR——5

Poa pratensis (June Grass; Kentucky Blue Grass; Spear Grass).

A perennial grass, growing usually 1½ to 2 feet high, with an abundance of long, soft, radical leaves, and sending off numerous running shoots from the base. The panicle is pyramidal or oblong in outline, from 2 to 4 inches long, the branches mostly in fives, at least below, 1 to 2 inches long, open and spreading, the longer ones flowering above the middle. The spikelets are about 2 lines long, ovate, closely three to five-flowered, mostly on very short pedicels. The outer glumes are acute; the flowering ones acute or acutish, five-nerved, the lateral nerves prominent, the lower part of the lateral nerves and of the keel more or less hairy, and the base clothed with webby hairs.

There are several well-marked varieties, which are much modified and improved by cultivation. It is indigenous in the mountainous regions of this country as well as of Europe, and has been introduced into cultivation in many countries.

Its principal use is as a pasture grass and for lawns. For hay-making there are many other grasses which furnish a heavier and more profitable crop. It is a grass which seems to require special conditions to bring out its best qualities, and hence it is held in very light or very great estimation in different regions. In England it is used but little, and never sown alone, but is generally recommended as a constituent of permanent pastures because of the earliness of its growth. In New Zealand, where it has been introduced, it is considered a curse rather than a blessing, because it overruns alike pastures and cultivated ground, and is as difficult of extermination as quack grass (*Agropyrum repens*). It varies much in size and appearance according to the soil in which it grows.

In all the Middle and Eastern States it forms the principal constituent of pastures, but in some parts it is not highly esteemed. From the unexampled success its cultivation has met with in Kentucky it has acquired the name of Kentucky blue grass.

The following very valuable notes on this grass are from the pen of Major Alvord, in Cassell, Peter & Co.'s work on Dairy Farming:

The *Poa pratensis* of the botanist has obtained a very wide reputation as the Kentucky blue grass, and led many into the mistaken belief that it was a peculiarly American grass, confined to the famous pastures of the region whence it derived its name. On the contrary, it is one of the most common grasses in nearly all parts of the country, being variably known as June grass, green meadow grass, common spear grass, and Rhode Island bent grass, and it is the well-known smooth-stalked meadow grass, or greensward, of England. There is no grass that accommodates itself to any given locality with greater facility, whether it be the Mississippi Valley, New England, Canada, the shores of the Mediterranean, or the north of Russia. It is found thriving upon gravelly soils, alluvial bottoms, and stiff clay lands in the permanent pastures of Missouri, and along the roadsides of Minnesota. Soil and climate cause varieties in its size and appearance, and this protean habit accounts for the various names by which it is known.

It probably attains its highest luxuriance and perfection as a pasture grass in the far-famed blue grass district of Kentucky. The central part of Kentucky, an area of 15,000 square miles or more, over limestone foundation, seems to be the richest blue grass country. There its seed-stalks are 2 to 3 feet high, with several long, parallel-sided leaves to each plant, and radical leaves often numbering thirty to a stalk. The root is perennial and throws off numerous and long-creeping root-stocks, enabling it to form a dense matted tuft. The chief reputation of this grass is as a pasture grass; the sod is easily obtained and very enduring, there being no such thing known as its running out on good land. Pastures sixty years unbroken afford their owners an annual profit of at least $10 an acre. It starts very early in the spring, and grows rapidly after being grazed off. It will furnish more late feed than most grasses, and no amount of pasturing is sufficient to utterly destroy it. It endures the frosts of winter better than any other grass on the continent, and therefore pushes its way northward into the Arctic Circle. Severe droughts injure blue grass, yet it grows as far south as the hilly parts of Georgia and Alabama, and in Arkansas, not, however, as vigorously as farther north. Although in a drought it often becomes dry enough to burn, it is greedily eaten by stock; it dries full of nourishing properties, and cattle will fatten upon it unless it has been drenched with rains. Blue grass can not be recommended for the meadow, as it is hard to cut and difficult to cure; the foliage is too short and too light after being dried.

It is an excellent grass for lawns, as it makes a dense, uniform mat of verdure, and sends up but one flowering stem a year; for this purpose it is thickly seeded and and kept closely mown.

An experienced Kentucky agriculturist says the season of sowing may be any time from August to April.

The seed should be sown from 1½ to 2½ bushels per acre, and lightly brushed in on a well-prepared surface. The seed may be sown on a grain field without any preparation. Some prefer to sow on small grain in February or March, on the snow. One advantage in this is the evenness with which the seed may be sown. If the sowing is done later it would be advisable to harrow the field before sowing it, and roll it afterward. A very loose or open surface is fatal to blue grass in the young state if the weather be the least dry. No stock should be permitted on the grass the first year. Blue grass is sometimes destroyed in sandy soils by cattle, which in grazing pull it up. In stiff clay this is not so likely to happen.

(Plate 75.)

Poa serotina (Fowl Meadow Grass).

Culms erect, 2 or 3 feet high, without running rootstocks. The leaves are narrowly linear, 3 to 6 inches long, and 2 to 3 lines wide, the sheaths long, smooth, and striate, the ligules long. The panicle varies with the size of the plant, from 5 to 10 or 12 inches long and 1 to 3 inches wide and lax; the branches mostly in fives or more numerous, nearly erect, from 1 to 4 inches long, the longer ones subdivided and flowering above the middle. There are some mountain forms or varieties in which the culms are 1 foot or less in height and the panicle greatly reduced. The spikelets are 1 to 2 lines long, two to five-flowered, on short pedicels. The outer glumes are about 1 line long and sharp-pointed. The flowering glume is rather obtuse, the lateral nerves not prominent, slightly pubescent on the margins below, and somewhat webby at the base.

This species is most common in the Northern States, particularly in New England, New York, and westward to Wisconsin, and also in reduced forms in all mountainous districts.

Professor Beal says:

The name fowl meadow grass is said to have been applied to this grass because ducks and other wild water-birds were supposed to have introduced the grass into a poor, low meadow in Dedham, Mass.

Mr. J. S. Gould, of New York, says:

I have found it to grow on almost every kind of soil, but it attains the greatest perfection in a rich, moist one. It is one of those grasses that thrive best when combined with others; it will not make a superior turf of itself, but it adds much to the value of a sward from its nutritive qualities and powers of early and late growth. As it perfects an abundance of seed it may be easily propagated.

Professor Phares, of Mississippi, says:

In portions of the Western States the grass has for some years been very highly recommended. In the Eastern States it has been cultivated for one hundred and fifty years or longer and valued highly. Jared Elliott, in 1749, spoke of it as growing tall and thick, making a more soft and pliable hay than Timothy and better adapted for pressing and shipping for use of horses on shipboard. He says it makes a thick abundant growth on land more moist than is adapted to common upland grasses, and may be mowed any time from June to October, as it never becomes so coarse and hard, but the stalk is sweet and tender and eaten without waste. It has not been sufficiently cultivated in the Southern States, so far as I am aware, to know how long a meadow set with it may remain profitable. It is, however, worthy of extended trial.

Mr. Charles L. Flint says:

It grows abundantly in almost every part of New England, especially where it has been introduced and cultivated in suitable ground, such as the borders of rivers and intervals occasionally flooded. It never grows so coarse or hard but that the stalk is sweet and tender, and eaten without waste. It is easily made into hay, and is a nutritive and valuable grass.

(Plate 76.)

Poa tenuifolia.

This species, in several varieties, is common in California, Oregon, Montana, etc., and is one of the numerous bunch grasses referred to in the accounts of the wild pasturage of that country. The foliage of some forms is scanty, but of others the radical leaves are long and abundant. It is stated that the Indians gather its seeds for food. (Plate 77.)

Poa trivialis (Rough-stalked-Meadow Grass).

This species very much resembles the *Poa pratensis.* It is distinguished chiefly by its having rough sheaths, by its long, pointed ligules, its fibrous roots, and the smooth, marginal nerves of the flowering glumes, whereas in *Poa pratensis* the sheaths are smooth, the ligules obtuse, the root stock running, and the marginal nerves of the flowering glumes are hairy.

It has been little cultivated by itself in this country, but is sometimes found in low meadows or on the banks of shaded streams. It flourishes best in low or wet ground and in shaded situations, and is not so well adapted to general cultivation as the blue grass.

Professor Phares, of Mississippi, says:

It is especially adapted to wood pastures, as it delights in shade, banks of streams, and moist ground generally. It bears tramping, and is an excellent pasture grass. It makes a good mixture with redtop and tall oat grass, and with other pasture grasses.

Poa trivialis var. occidentalis:

This grass, apparently a variety of *Poa trivialis*, appears to be indigenous in Colorado and New Mexico. It has a larger, looser panicle than the introduced plant.

Poa andina.

This is a smooth, rigid, perennial grass, growing on the great western plains in arid situations. It varies in height from 1 to 2 feet, with short, rigid, pointed root-leaves, and with usually one or two stem-leaves, the upper one with a very short blade, or almost none. The panicle is close and rather dense, 2 to 3 inches long, the spikelets about three-flowered, the empty glumes rather large and broad, and the flowering glumes pubescent on the nerves below.

It is probable that this species may be introduced with advantage into cultivation in the arid districts of the West. (Plate 78.)

GLYCERIA.

Spikelets terete or flattish, several to many-flowered, in a narrow or diffuse panicle, the rhachis smooth, and readily disarticulating between the flowers; outer glumes shorter than the flowers, unequal, membranaceous, one to three-nerved, unawned; flowering glume membranaceous to subcoriaceous, obtuse, awnless, more or less hyaline and denticulate at the apex, rounded (never keeled) on the back, five to nine-nerved, the nerves separate, and all vanishing before reaching the apex; palet about as long as its glume, two-keeled, entire or bifid at the apex.

The species of this genus are seldom employed in cultivation. They mostly grow in wet or swampy ground, and where found in abundance can be utilized for pasturage or hay-making.

Glyceria arundinacea (Tall Meadow Grass; Reed Meadow Grass).

This species is widely diffused in the northern portions of the United States and Canada, and in the Rocky Mountains from Mexico to Montana. It has a stout, erect, leafy culm, 3 to 4 feet high. The leaves are a foot or two long, a quarter to half an inch wide, flat, and somewhat tough on the edges. The panicle is large, 9 to 15 inches long, and much branched. (Plate 79.)

Glyceria Canadensis (Rattlesnake Grass; Tall Quaking Grass).

The culms stout, about 3 feet high, smooth and leafy. The leaves linear-lanceolate, 6 to 9 inches long, or the lower ones much longer, about 4 lines broad and rather rigid. The panicle large and effuse, 6 to 9 inches long, oblong, pyramidal, and at length drooping. The whorls an inch or more distant, the branches semi-verticillate, mostly in threes, the largest 3 to 4 inches long, and subdivided from near the base.

A grass belonging to the northern portion of the United States, usually found in mountainous districts, in swamps, and on river borders, growing in clumps. It is quite an ornamental grass, resembling the quaking grass (*Briza*). Cattle are fond of it, both green and when made into hay. It is well adapted to low meadows.

Glyceria fluitans (Floating Manna Grass).

Culms are usually 3 to 4 feet high, rather thick and succulent, and quite leafy. The leaves are 4 to 9 inches long, and 3 to 4 lines wide. The panicle is often a foot long, very narrow, the short distant branches mostly in twos or threes, 1 or 2 inches long, erect and close, each having usually two or four spikelets. The spikelets are half an inch to three-quarters of an inch in length, rather cylindrical and nearly of the same thickness throughout, seven to thirteen-flowered.

This species grows in shallow water on the margins of lakes, ponds, and sluggish streams.

Hon. J. S. Gould says:

This grass is found growing in shallow water, overflowed meadows, and wet woods, but will bear cultivation on moderately dry grounds. Schreber says that it is cultivated in several parts of Germany, for the sake of the seeds, which form the manna crop of the shops, and are considered a great delicacy in soups and gruels. When ground into meal they make bread, very little inferior to that made from wheat. In Poland large quantities of the seed are obtained for culinary purposes. All granivorous birds are exceedingly fond of these seeds. Trout, and indeed most fish, are very fond of them; wherever it grows over the banks of streams the trout are always found in great numbers waiting to catch every seed that falls.

There is a great difference of opinion among agricultural writers with respect to the fondness of animals for the leaves and culms of this grass. We have often seen the ends of the leaves cropped by cattle, but have never seen the culms or root-leaves touched by them. On the other hand, reliable writers have asserted that cattle, horses, and swine were alike fond of it.

Glyceria nervata (Nerved Meadow Grass).

This is similar in appearance to the tall meadow grass, but is smaller, with a lighter panicle and smaller flowers. It has also much the same general range. It usually grows along the wet margins of streams and swamps. It is nutritious and might be advantageously mixed with other grasses in wet grounds. It is especially abundant in the Rocky Mountains. It is sometimes improperly called fowl meadow grass. No attention has been given to its cultivation in this country. In the Woburn Agricultural Experiments conducted in England by the Duke of Bedford, this grass was under trial, and was very highly esteemed. Mr. Sinclair states that in February, 1814, after the severe winter preceding, this grass was perfectly green and succulent, while not one species of grass, out of nearly three hundred that grew around it remained in a healthy state, but were all inferior and more or less injured by the severity of the weather. The aftermath was found to be remarkably abundant and nutritive. It was found to be adapted to most soils except such as were tenacious. Mr. Sinclair also said that further experience in the cultivation of the grass enabled him to state that it possesses very valuable properties, and that it will be found a valuable ingredient in permanent pastures, where the soil is not too dry, but of a medium quality as to moisture and dryness. (Plate 80.)

FESTUCA.

Spikelets three to many-flowered, variously panicled, pedicellate, rhachis of the spikelets not hairy; outer glumes unequal, shorter than the flowers, the lower one-nerved, and the upper three-nerved, narrow, keeled, acute; flowering glume membranaceous, chartaceous, or subcoriaceous, narrow, rounded on the back (not keeled), more or less distinctly three to five-nerved, acute or tapering into a straight awn, rarely obtusish; palet narrow, flat, prominently two-nerved or two-keeled.

Festuca elatior (Meadow Fescue Grass; Tall Fescue; Randall Grass).

A perennial grass, growing from 2 to 4 feet high, with flat, broadish leaves about a foot long. The panicle is somewhat one-sided, loose, and spreading when in flower, contracted after flowering, from 6 to 10 inches long, the branches 1 to 2 inches long, erect, mostly in pairs below, single above, subdivided; the spikelets are lanceolate or linear, about half an inch long, five to ten-flowered. The flowering glume is lanceolate, about three lines long, firm in texture, five-nerved, scarious at the margin, acute, and sometimes with a short but distinct awn at the apex.

This is an introduced species now frequently met with in meadows; it is one of the standard meadow grasses of Europe. Cattle are said to be very fond of it, both green and as hay.

There is a smaller form or variety, which is the variety *pratensis* or *Festuca pratensis*, Hudson.

Professor Killebrew, of Tennessee, writes of it as follows:

This grass has received some attention in different parts of the State, and has met with a warm reception from those testing it. It ripens its seeds long before any other grass, and consequently affords a very early nip to cattle. It has been raised under various names in Virginia, as "Randall grass," and in North Carolina as "evergreen grass."

Mr. James Taylor, writing from North Carolina, says:

The evergreen grass is very good for pasturing through the fall and winter. It will do best when sown on dry land, and is well adapted to sheep. It grows well on rocky soil to the height of 4 or 5 feet when ripe, continuing green in the spring, and affording fine herbage throughout the winter. It is best to sow in the spring, with oats. A peck of well-cleaned seed is enough for an acre, or a bushel in the chaff. It ripens about the first of June. If sown in the spring this grass will not go to seed before the next year, but if sown in the fall it will bring seed the next spring. From the limited cultivation it has met with in Tennessee, it seems to be better adapted to moist, low lands, though I have seen it growing on some of the high ridges of East Tennessee, at least 1,500 feet above the sea. There it thrives luxuriantly, and makes a very superior pasture.

Professor Phares, of Mississippi, says:

It grows well in nearly all situations, wet or dry, on hill or bottom land, even though subject to overflow, and matures an extraordinary quantity of seed. The seeds germinate readily, and it is easy to set a piece of land with this grass. Seeded alone, 23 pounds, or about 2 bushels of seed, should be sown broadcast in August, September, October, or from the middle of February to the 1st of April. From remaining green through the winter it is sometimes called "evergreen grass." Mowed and dried it makes a good hay, much relished by stock.

(Plate 81.)

72

Festuca ovina (Sheeps' Fescue).

A densely tufted, perennial grass, with an abundance of rather narrow, sometimes involute, short, radical leaves, and slender culms, 1 to 1½ feet high. The panicle is 2 to 4 inches long, narrow, the branches mostly single and alternate, erect and few-flowered; the spikelets are mostly three to five-flowered, and about 3 lines long; the outer glumes are acute and narrow. The flowering glumes are lanceolate, two lines long, roughish, and with a short, rough awn about half a line long.

This species has many varieties both in this country and in Europe. It is indigenous in the mountainous parts of New England, in the Rocky Mountains, and in various northern localities.

As found in cultivation it has been derived from Europe.

Hon. J. S. Gould, of New York, says:

It forms the great bulk of the sheep pastures of the highlands of Scotland, where it is the favorite food of the sheep, and where the shepherds believe it to be more nutritious for their flocks than any other. Gmelin says that the Tartars choose to encamp during the summer where this grass is most abundant, because they believe that it affords the most wholesome food for all cattle, but especially for sheep. Nature distributes it among dry, sandy, and rocky soils, where scarcely any other species would grow. It is without doubt the very best of the grasses growing on sandy soils. It roots deeply, and forms a dense, short turf, which adapts it admirably for lawns and pleasure grounds, where the soil is sandy. It is almost useless as a hay crop, as its leaves and culms are too fine to give a remunerative amount of hay ; it is only as a pasture grass on sandy soils that it is valuable ; and in these, when highly manured, it is driven out by the more succulent species. It is often found 4,000 feet above the level of the sea. Its seeds weigh about 14 pounds to the bushel.

(Plate 82.)

Festuca scabrella (Bunch Grass).

The culms are usually 2 to 3 feet high, erect, and smooth; the radical leaves are numerous, about half as long as the culms, generally rigid, involute, and scabrous on the margins; the blade is prone to separate when old, leaving an abundance of leaf-less sheaths at the base ; the cauline leaves are about two, short and pointed, 2 to 4 inches long ; the sheaths scabrous, the ligule short or wanting ; the panicle is usually 3 to 5 inches long.

A perennial grass growing in strong clumps or bunches, and hence called "bunch grass." It is a native of the Rocky Mountain region, from Colorado westward to California and Oregon.

In Montana it is called the great bunch grass and is one of the principal grasses of that country. It is the prevailing species on the foot-hills and mountain slopes at from 6,000 to 7,000 feet altitude. "It is rather too hard a grass for sheep, but there is no grass more valued on the 'summer ranges' for cattle and horses. It makes excellent hay for horses and is cut in large quantities for this purpose. It grows in large tussocks, making it rather a difficult grass to mow with a machine." It is one of the most important grasses of eastern Oregon and Washington. (Plate 83.)

BROMUS.

(BROME GRASS.)

Spikelets five to many-flowered, in a dense, or lax, or diffuse panicle; the rhachis between the flowers glabrous; outer glumes more or less unequal, shorter than the lowest flower, membranaceous, acute, awnless, or short mucronate, one to nine-nerved; flowering glume from membranaceous to rigid, and subcoriaceous, rounded on the back or compressed and keeled, five to nine-nerved, acute, and awned from below the mostly two-cleft apex; palet rather shorter than the glumes, two-keeled, the keels rigid and ciliate; grain adhering to the palet.

Bromus secalinus (Chess; Cheat).

It is an old tradition which some farmers still cling to that chess is a degenerated wheat; that the action of frost and other causes occasion the deterioration, whereas the truth undoubtedly is that chess seed was either in the land or in the seed sown, and, being more hardy than wheat, it survived the frost and took possession of the ground. Some years ago this grass had a temporary popularity under the name of Willard's brome grass, but it was soon abandoned when brought into competition with better grasses.

In the South it would perhaps be a good winter grass, like its relative *Bromus unioloides*, but it is not as vigorous a grass as that species, and does not produce such an abundance of foliage. (Plate 84.)

Bromus unioloides (Schrader's Grass; Rescue Grass).

In its early growth it spreads and produces a large amount of leaves; early in the spring it sends up its flower stalks, which grow about 3 feet high, with a large, open, spreading panicle, the ends of the branchlets bearing the large, flattened spike-lets, which, when mature, hang gracefully upon their stems, giving them quite an ornamental appearance. These spikelets are from 1 inch to 1½ inches in length, and composed of two acute, lanceolate glumes at the base, and from seven to ten flowers, arranged in two rows alternate on each side of the axis. The flowers are lanceolate, or ovate-lanceolate, the flowering glume extending into a fine point or short awn.

This is one of the so-called winter grasses; that is, it makes, in the South, a large share of its growth during the winter months.

During several years past this grass has been sent to the Department, chiefly from Louisiana and Texas, and has been much commended. Many years since the same grass was distributed and experimented with under the name of Australian oats, or *Bromus Schraderi.* It is not adapted to use in a country with severe winters, and hence did not give satisfaction in all places.

Mr. C. Mohr, of Mobile, says of it:

Only of late years found spreading in different parts of this State; makes its appearance in February, grows in tufts, its numerous leafy stems growing from 2 to 3 feet high; it ripens the seed in May; affords in the earlier months of spring a much-relished, nutritious food, as well as good hay.

It is said to have been introduced into Georgia by General Iverson, of Columbus, and by him called rescue grass. The favorable opinion which it at first received does not seem to have been well sustained in that State.

Professor Phares, of Mississippi, says:

This grass is also called *Bromus Schraderi, Bromus Willdenovii, Ceratochloa unioloides,* and *Festuca unioloides.* It is an annual winter grass. It varies in the time of starting growth. I have seen it ready for mowing the first of October, and furnish frequent cuttings till April. Again, it may not start before January nor be ready to cut till February. This depends on the moisture and depression of temperature of the fall, the seeds germinating only at a low temperature. When once started, its growth after the successive cuttings or grazings is very rapid. It is tender, very sweet, and stock eat it greedily. It makes also a good hay. It produces an immense quantity of leaves. On loose soil some of it may be pulled up by animals grazing it.

(Plate 85.)

Bromus ciliatus.

A tall, coarse species, much addicted to rocky woodlands, but of no agricultural value.

LOLIUM.

Spikelets several-flowered, solitary on each joint of the continuous rhachis of the simple spike, placed edgewise against the rhachis, the glume wanting on the inside, the outer empty glume nearly as long as, or longer than, the spikelets; flowering glume rounded on the back, not keeled ; palet shorter, two-keeled.

Lolium perenne (Italian Rye Grass).

A perennial grass, introduced from Europe. The culms are 2 to 3 feet high, very leafy, and terminating in a loose, spike-like panicle, 6 inches or more in length. The spikelets are arranged alternately on the axis, placed edgewise ; that is, with one edge of the flat spikelet applied to the main stem at short distances, so that there may be twenty or more in the panicle. The spikelets are one-half to three-fourths of an inch long; generally seven to eleven-flowered. The inner empty glume is generally wanting; so that, except on the terminal spikelets, only one glume is apparent, which is half or more than half the length of the spikelet, narrowly lanceolate, and acute. The general appearance of the panicle is like that of couch grass (*Agropyrum repens*). The flowering glumes are thickish, obscurely nerved, rather hispid, acutely pointed, or, in the variety *Italicum,* with a rather long awn. The proper palets are similar to the flowering glumes, and of nearly equal length.

An intelligent writer whom we have frequently quoted, says, respecting this grass :

It occupies the same place in Great Britain that Timothy does with us, and is there esteemed, on the whole, higher than any other species of grass, and is called rye grass or ray grass. Of all the varieties of *Lolium perenne* which are known, that called *Italicum* is by far the most valuable. Its spikelets are conspicuously bearded, the flowers being all terminated by long, slender awns, which character distinguishes it very easily from *Lolium perenne.* Its name (Italian rye grass) is derived from the fact that its native habitat is on the plains of Lombardy, where broad and extensive plains of pasture land are frequently inundated by the mountain streams which intersect them. It is mainly adapted to irrigated meadows, and in these it is undoubtedly superior to any other grass.

Professor Phares says:

This grass stands drought well and grows most luxuriantly in our Southern States. If not kept grazed or mowed, however, the leaves cover the ground so deeply and densely that an excess of rain in very hot weather in the extreme South causes it to rot suddenly, destroying even the roots. This I have never seen or heard mentioned by any other person, but it occurred on my own farm one season, where I was reserving a lot for seed.

(Plate 86.)

Lolium temulentum (Poison Darnel).

This species is frequently found in grain fields. The seeds have long enjoyed a reputation of being poisonous to stock, and also to mankind when mixed in large quantity with the wheat or rye used in the making of bread. The question seems hardly yet decided, but it is best to exterminate the grass as a weed and a pest.

AGROPYRUM.

Spikelets several-flowered (three to nine, or more), compressed, alternately sessile on the continuous or slightly-notched rhachis of the simple spike, and with the side against the rhachis; outer glumes nearly equal and opposite, membranaceous or herbaceous, one to three-nerved, scarcely keeled, tapering to a point or awned; the flowering similar to the outer ones, rounded on the back; three to seven-nerved, pointed or awned from the apex; palet nearly as long as its glume, the two prominent nerves almost marginal, scabrous ciliate.

Agropyrum glaucum (Blue Stem ; Bluejoint).

This species, which has been considered a variety of the next, prevails on the Western plains from Texas to Montana, and is well known to stockmen. It differs from *Agropyrum repens* in having a stiffer, more erect and rigid stem and leaves, the leaves often becoming involute. It is generally of a light, bluish-green color. The spike is generally shorter, denser, and with larger spikelets.

Professor Scribner, writing of this grass in Montana, says:

It is the most highly praised of the native grasses for hay. Wherever it occupies exclusively any large area of ground, as it does frequently in the lower districts, especialy near Fort Benton, it is cut for hay. Naturally it does not yield a great bulk, but its quality is unsurpassed. After two or three cuttings the yield of hay diminishes so much that it is scarcely worth the harvesting. It is then customary to drag a short-toothed harrow over the sod, which breaks up the creeping roots or underground stems, and each fragment then makes a new plant.

The same valuable opinion of this grass is entertained by stockmen in Nebraska, Colorado, and New Mexico. It occurs nearly everywhere, but sparsely, on the plains, and extending quite up into the mountains. In the valleys and along streams it frequently forms large patches and grows closer and more abundant, when it is commonly cut for winter use. (Plate 87.)

Agropyrum repens (Couch Grass; Quack Grass).

There has been a good deal of discussion relative to this grass, some pronouncing it one of the vilest of weeds, and others claiming for it high nutritive qualities overweighing all the disadvantages of its growth. Whichever party may be right, it is proper that farmers should be acquainted with it in order to know how to treat it, and hence our description. It forms a dense sod by means of its far-reaching rhizomas or root stocks, which have short joints, and roots tenaciously at every joint.

It has an abundance of foliage, and sends up a flowering culm 2 to 3 feet high, which is terminated by a close, narrow spike of flowers from 3 to 6 inches long. This spike consists of a succession of closely set spikelets, one at each joint of the axis, and placed flatwise with the side against the stalk. Each spikelet contains several (three to eight) flowers, with a pair of nearly equal and opposite three to five-nerved glumes at the base.

Hon. J. S. Gould says:

The farmers of the United States unite in one continuous howl of execration against this grass, and it seems strange, when every man's hand is against it, that it is not exterminated. Yet, we could never really satisfy ourselves that its presence in meadows and pastures was such an unmitigated curse. In lands where alternate husbandry is practiced it must be admitted to be an evil of great magnitude. Its hardiness is such, and its rapidity of growth is so great, that it springs up much more rapidly than any other crop that can be planted, and chokes it. Still, it has many virtues. It is perfectly cosmopolitan in its habits. It is found in all sorts of soil and climates. Its creeping roots are succulent and very nutritive, and are greedily devoured by horses and cows.

(Plate 88.)

Agropyrum tenerum.

This grass prevails in the Rocky Mountain region from New Mexico to Oregon, and has been commonly called a variety of *Agropyrum repens*, from which it differs essentially in wanting the running root-stalks, in a narrower, nearly cylindrical spike, and in growing in clumps. It occurs mostly in low, moist grounds, and, like the *Agropyrum glaucum*, it is one of the best grasses for hay. It ripens in July, and affords very little feed thereafter.

HORDEUM.

Inflorescence a dense spike, with two or three spikelets at each joint of the notched rhachis; spikelets one-flowered, with an awl-shaped rudiment of a second flower, the central spikelet of the cluster perfect and sessile, the lateral ones short-stalked and imperfect or abortive; outer glumes side by side, two to each spikelet, usually slender and awn-pointed, or bristle form; flowering glume herbaceous, shorter, oblong, or lanceolate, rounded on the back, not keeled, five-nerved, acute or long-awned; palet shorter, two-keeled.

Hordeum jubatum (Wild Barley; Squirrel-tail Grass).

On the sea-coast and saline soil in the interior, especially on the Rocky Mountains. It has no agricultural value, but its long-barbed awns are injurious to the mouths of cattle.

Hordeum murinum.

Professor Brewer states that this grass, unfortunately, is extensively naturalized in California and is a vile pest; it comes in when land is overstocked; is known there as "squirrel grass," "squirrel tail," "foxtail," and "white oats." The heads break up and the barbed seeds work into the wool of sheep and even into the flesh of lambs, killing them. It damages the eyes and throats of animals.

Hordeum pratense.

An annual or biennial grass growing principally in alkaline soil in the Western States and Territories. It is eaten by cattle when in a young state, but when mature it is worthless and pestiferous on account of its barbed awns.

ELYMUS.

Spikelots two to four at each joint of the rhachis of the simple stout spike, sessile, one to six-flowered ; outer glumes two for each ; spikelets nearly side by side in its front, forming a kind of involucre for the cluster, narrow, rigid, one to three-nerved, acuminate or awned ; flowering glumes herbaceous, rather shorter, oblong or lanceolate, rounded on the back, not keeled, acute or awned ; palet shorter than its glume, two-keeled.

Elymus Canadensis (Wild Rye ; Rye Grass ; Lyme Grass).

A perennial, coarse grass, growing on river banks and in rich, shaded woods. In some localities, especially on moist prairies and banks in the west, it is quite common and is cut for hay. It should be cut early to be of value. (Plate 89.)

Elymus condensatus (Giant Rye Grass.)

This is a perennial grass, ranging from San Diego throughout California, and into Oregon and Washington Territory, also in the Rocky Mountain region of the interior. It is very variable, but always a strong, heavy-rooted, coarse grass, from 3 to 5 or even to 12 feet high. Mr. Bolander states that it seems to do excellent service by fixing the soil on the banks of creeks and rivers. In the larger forms the culms are half an inch thick. The leaves are smooth, 2 feet long and an inch wide or more, and the panicle 8 to 14 inches long and 1½ inches thick. As it usually occurs in arid grounds, it is from 3 to 6 feet high, the leaves about 1 foot long and half an inch wide, and the spike-like panicle 4 to 8 inches. In the large form the branches of the panicle are subdivided and 1 or 2 inches long.

Mr. W. C. Cusick, of Oregon, says:

This is a very valuable grass, commonly known as rye grass. In Baker County large quantities are cut for hay, for which it is said to be excellent. It is also much used as a winter forage plant. Cattle are driven into the dry bottoms, where it grows, and live upon it when the shorter grasses are covered with snow.

(Plate 90.)

Elymus triticoides.

This has been considered a variety of *Elymus condensatus*, from which it differs in having strong runners, and not growing in thick clumps, but scattering and singly. Mr. Cusick says it is a valuable grass in Oregon, and cut for hay in wild meadows.

Elymus Virginicus (Wild Rye Grass; Terrell Grass).

The culm is rather stout, 2 to 3 feet high, leafy ; the lower leaves are 10 to 15 inches long, broad and rough. The sheath of the upper leaf usually incloses the stalk, and sometimes the base of the flower-spike. This spike is erect, dense, and rigid, 2 to 4 or 5 inches long, and one-half inch thick. The spikelets are two or three together at each joint, all alike and fertile, sessile, two to five-flowered, and each with a pair of empty glumes. These glumes are very thick and coarse, strongly nerved, lanceolate and bristle-pointed, about 1 inch long. The flowering glumes terminate in a stiff, straight awn, half an inch to nearly an inch long, the lowest one in the spikelets having the longest awn, the others gradually shorter. The palet is oblong, obtuse, and as long as the flowering glume, excluding the awn.

A coarse, perennial grass, growing on alluvial river banks, or in rich, low grounds. This grass frequently forms a considerable portion of native meadow lands, and makes a coarse hay. It starts growth early in the spring, and thus affords a good pasturage. Professor Killebrew, of Tennessee, says it is very valuable and ought to be tried in cultivation.

Professor Phares, of Mississippi, says:

This perennial grass is a native of the Southern States. As all farm stock, except hogs, are fond of it, and it is green through the winter and spring, it has been destroyed when grazing animals have access to it at all times. It is, however, found in many of our States, along the banks of wooded streams, of ditches, and in fence corners among briers and thickets. It will grow on thin clay, gravelly, or sandy soil, but much better on rich lands, dry or rather moist, and will thrive ten, twenty, or more years on the same land.

(Plate 91.)

ARUNDINARIA.

Spikelets many-flowered, flattened, racemose or paniculate, the uppermost flowers imperfect; outer glumes very small, membranaceous, the upper one larger; flowering glumes much larger, membranaceo-herbaceous, convex on the back, not keeled, many-nerved, acuminate, mucronate, or bristle-pointed; palet shorter than its glume, prominently two-keeled.

Arundinaria tecta (Switch Cane; Small Cane).

Professor Phares, of Mississippi, says of this grass :

This largest of our grasses has a hard, woody stem from one-half to 3 inches in diameter, and from 10 to 40 feet high, erect, tapering from near the base, jointed every 8 to 12 inches for one-half the length or more, then the joints becoming shorter and smaller to the top; leaves 1 to 2 inches wide, persistent, on clustered, spreading branches which also are jointed and appear the second year. On rich land in spring the young stems shoot up full size, ten or twenty feet high, and are as crisp as asparagus, and by some persons as much relished. Hogs, cattle, and other animals are fond of the young plants and seeds. The age at which the large cane blooms has not been definitely decided. It probably varies with the latitude, soil, and surroundings, from ten to thirty years. When the seeds mature the cane dies. Grazing animals feed greedily on the leaves in the winter and find protection from the driving rains and piercing winds under the dense roof of the canebrake or thicket. The stems are used for fishing-rods, scaffolds for drying cotton, for pipe-stems and pipes, and splints for baskets, mats, and other purposes. The small cane is different in habit from the large cane. It blooms sometimes two or more consecutive years without dying down to the root. Live stock like it as well as the large cane. Both grow best on rich lands, hills, or bottoms; but they will grow on thin clay soil, improve it, and if protected from stock, rapidly extend by sending out long roots (rootstocks) with buds.

The small cane is found sparingly as far north as Baltimore, Md. The large cane is probably confined to the Gulf States, but this is uncertain.

CULTIVATED FORAGE PLANTS OTHER THAN GRASSES.

Order LEGUMINOSÆ.

The CLOVER FAMILY.

This order is characterized by having alternate, usually compound, leaves, with stipules; flowers polypetalous, the calyx mostly five-lobed, the corolla generally with five irregular petals, usually ten stamens, sometimes five, or many, usually united by the filaments, or nine united and one free, or sometimes all distinct; the ovary a one-celled carpel, becoming a legume or pod with few or many seeds, the pod sometimes marked into joints called *loments.*

The order embraces an immense number of plants of varying character, some small and insignificant, some trees of large size. Many of the most useful vegetable products are obtained from it.

TRIFOLIUM.

(THE CLOVERS.)

This genus is one of the most useful of the order and embraces a large number of species, several of which are well known in cultivation.

The genus is characterized by having the leaves mostly trifoliate; that is, made up of three leaflets at the end of the leaf-stalk; some species have five or more leaflets, either close together at the end of the leaf-stalk or somewhat scattered in opposite pairs. The flowers are collected in roundish or oblong heads, with or without a general involucre. The calyx is five-toothed, the petals five, irregular, persistent; nine stamens united and one free; the pod small, mostly inclosed in the calyx, and one to four-seeded.

Trifolium pratense (Red Clover; Common Clover).

Red clover is so well known to the agricultural community that it requires very little description. It is usually a perennial of a few years duration, a native of Europe and Asia, but early introduced into this country. Its cultivation is said to have begun in England about two hundred and fifty years ago. It is one of the most important of cultivated crops, both for feed for animals and as an improver of the soil.

A writer in the Country Gentleman says:

No matter how mismanaged, clover is a benefit, and whatever else he may do, the farmer who grows clover is making his farm better. It does not need cultivating; the long deep-reaching roots mellow and pulverize the soil as nothing else can. If it grows thriftily the top acts as a mulch, seeding the ground and keeping it moist. A crop of 2 tons or more of clover plowed under or cut for hay can hardly fail to leave the ground better than it was before. It should be the farmer's aim to grow the largest possible crop of clover.

79

The Rural New Yorker says:

Ten acres of good clover are worth more than so much wheat, if the value of what is left in the ground by the clover is taken into account. When a crop of wheat is taken the ground is exhausted of so much of its fertility, which is carried off in the wheat, but when a crop of clover is taken the soil is actually in better condition than before, and is good enough to yield a crop of wheat or corn.

A Wisconsin farmer says:

If you want to clear your land of weeds, sow clover and sow it thick. If you want to grow big corn-crops, grow clover and pasture off with hogs. Plow up the land in the fall, and the corn-crop following will make you happy. If you want to make rich farms and make money, grow clover, corn, and hogs.

Professor Beal says:

Red clover is well adapted to many portions of the temperate regions of the earth. It likes best a soil of clay loam, rich in lime, but will thrive better than Timothy and most other true grasses where the land is sandy or gravelly. On good grass-land it is usually the custom to sow Timothy with red clover, although it blossoms some three weeks later. Many prefer to sow orchard grass with clover, as they flower and are ready to cut at the same time. Timothy is well adapted to sow with the large, late, or mammoth clover.

There are some portions of the country where, owing either to climate or soil, red clover has not been successful, and in those places some other leguminous plant can generally be substituted with advantage.

Trifolium medium (Mammoth Clover).

The true botanical position of the clovers cultivated in this country under the names of mammoth, sapling, or pea-vine clover, etc., is still somewhat in doubt. They are usually regarded as being the above-mentioned species, but are perhaps a variety or varieties of the common red clover, *Trifolium pratense.*

They agree in having a larger and later growth than the ordinary red clover, and on this account are for some purposes more valuable. The following records of experience may be relied upon for the localities mentioned.

Prof. Samuel Johnson, Agricultural College, Michigan:

It grows too rank and coarse to make good hay. For pasture or for manurial purposes it might prove better than the smaller sort. When grown for seed it is usually pastured until the 1st of June, and then allowed to grow up and mature the crops.

M. O. Alger, Augusta, Michigan:

Pasturing until the first of June insures a larger yield of seed, as it is cooler while filling, but many do not pasture. I do not think it can be cut more years than the smaller kind. It is said to stand drought better, but I doubt that. It will give three times the amount of pasture during the season that is given by the smaller kind if kept down pretty close, but during the fall the amount of pasture produced is less. It is said to smother out in winter if a large amount is left on the ground. Another objection is that it requires cutting just at harvest-time.

C. M. Alger, Newaygo, Michigan:

I have raised the mammoth clover, but do not like it for my heavy land, as it grows too large. For every acre that I raise I have to buy or borrow two more of my neighbor's to cure it on. It is, however, excellent for pasture, as it stays on the ground longer than the medium variety. It is good for raising seed, as it nearly always fills full. I have seen 8 bushels per acre. The seed is always grown on the first crop, as the second never blossoms. It grows here from 4 to 5 feet high and is good for plowing under for manure.

Austin Potts, Galesburgh, Michigan:

Perhaps not over 20 per cent. of the clover grown here is of the mammoth variety. It does not seed as well as the common clover.

L. H. Bursley, Jenisonville, Michigan:

I do not find it as good for hay as the common red clover; the stalks are so large that stock will not eat them at all. For pasture it is better than the small variety. It does not require pasturing in spring in order to produce a crop of seed.

James Hendricks, Albany, N. Y.:

About twenty years ago there was treble the quantity sown in this part of Albany County that there is at present; now nearly all our farmers sow the medium clover with Timothy.

Prof. F. A. Gully, Agricultural College, Mississippi:

On good land with us it grows rank, and the long stems fall down and mat on the ground, and if we happen to have wet weather the lower leaves and parts of the stalk will begin to decay before the plant is in full bloom.

The second crop ripens seed, but to what extent I can not say; I consider the common red clover more desirable here, although it may not yield as well.

Trifolium hybridum (Alsike Clover).

This differs from common red clover in being later, taller, more tender and succulent. The flower-heads are upon long peduncles, and are intermediate in size and color between those of white and red clover. The botanical name was so given from its being supposed by Linnæus to be a hybrid between those clovers, but it is now known to be a distinct species. It is found native over a large part of Europe, and was first cultivated in Sweden, deriving its common name from the village of Syke in that country. In 1834 it was taken to England, and in 1854 to Germany, where it is largely grown, not only for its excellent forage but also for its seed, which commands a high price. In France it is little grown as yet, and is frequently confounded with the less productive *Trifolium elegans*.

The following is condensed from "Les Prairies Artificielles," by Ed. Vianno, of Paris:

Alsike does not attain its full development under two or three years, and should therefore be mixed with some other plant for permanent meadows. It is best adapted to cool, damp, calcareous soil, and gives good results upon reclaimed marshes. It is adapted neither to very dry soils, nor to those where there is stagnant water. Being of slender growth, rye grass, rye, or oats are often sown with it when it is to be

3594 GR——6

mowed. In fertile ground weeds are apt to diminish the yield after a few years, so that it requires to be broken up. It is generally sown in May, at the rate of 6 or 7 pounds of the clean seed per acre. Sometimes it is sown in the pods at the rate of 50 to 100 pounds per acre, either in spring or in autumn after the cereals are harvested.

Alsike sprouts but little after cutting, and therefore produces but one crop and one pasturage. The yield of seed is usually 130 to 170 pounds per acre. The seed separates more easily from the pods than that of ordinary clover, and as the heads easily break off when dry, care is required in harvesting.

It does not endure drought as well as the common red clover, but will grow on more damp and heavy soils, and it is said that it can grow on land which, through long cultivation of the common clover, has become "clover sick." (Plate 92.)

Trifolium incarnatum (French Clover).

This annual clover is a native of Europe. It grows to the height of about 2 feet. The heads are about 2 inches long, very densely flowered, with the petals ranging from a pinkish to a crimson color.

It has been introduced and tried to some extent for cultivation in this country, but has not met with much favor. It deserves trial, however, in the dry climates of the West. (Plate 93.)

Trifolium repens (White Clover; Dutch Clover).

This is a small perennial species, with prostrate stems which take root strongly at the joints. It is said to be the shamrock of Ireland. It is a native of Europe and Northern Asia, and has been introduced into, and naturalized in, many other countries. It is said that, although indigenous in England, it only began to be cultivated at the beginning of the eighteenth century. On account of its creeping habit, when once established, it soon covers the ground and spreads extensively. Mr. Sutton, an English writer, says:

It prospers on mellow land containing lime, and on all soils rich in humus, from marl to gravelly clay. It does better in poor land than red clover. In early spring it produces very little food, and the plant is so dwarfed that it is practically useless for cutting for a crop of hay. Still, perennial white clover forms an essential constituent of every good pasture. All cattle eat it with relish, but it is of less use for the production of milk than of flesh, and is of special service in fattening sheep. It is not suitable for culture by itself, and its herbage is better for cattle when mingled with other grasses, especially with perennial rye grass.

A correspondent of Farm and Home says:

Every pasture should contain some white clover. It will afford more feed at certain times of the year than grass or any other kind of clover. It will not flourish in damp soils, or those that are very poor. It will do well in a partial shade, as a grove or orchard, but to make the highest excellence it should have the advantage of full sunlight. It is easy to secure patches of white clover in a pasture by scattering seed in early spring on bare places and brushing it in. One pound of seed is enough to start white clover in a hundred places. The disposition of this clover is to spread by means of the branches that run along the ground and take root.

Prof. W. J. Beal, of Michigan Agricultural College, says:

White clover is a fickle plant, coming and going with the varying seasons. It often burns out in hot weather. An old, hard road, once abandoned, is likely to send up white clover in advance of the grasses. It is a well-known and highly prized bee-plant. It is often sown with some of the finer grasses for lawns.

Trifolium stoloniferum (Running Buffalo Clover).

This is a native perennial species, growing about a foot high; long runners are sent out from the base, which are procumbent at first, becoming erect. The leaves are all at the base, except one pair at the upper part of the stem. The root-leaves are long-stalked, and have three thinnish obovate leaflets, which are minutely toothed. The pair of leaves on the stem have the stalk about as long as the leaflets, pointed and entire on the margins, the lower ones nearly an inch long, the upper ones about half as long. There are but one or two heads on each stem at the summit, each on a pedicel longer than the leaves. The heads are about an inch in diameter, rather loosely flowered, each flower being on a short, slender pedicel, or stem, which bends backward at maturity. Each flower has a long-toothed calyx about half as long as the corolla, which is white, tinged with purple.

*This species is found in rich open woodlands, and in prairies in Ohio, Illinois, Kentucky, and westward. It is of a very vigorous growth, but somewhat smaller in size than the common red clover. It should receive the attention of farmers and its value be ascertained by cultivation and experiment. (Plate 94.)

ONOBRYCHIS.

Onobrychis sativa (Sainfoin).

A perennial, having somewhat the appearance of Lucerne, but of smaller size and different habit. It seldom exceeds 1½ feet in height, with a weak stem, rather long, pinnate leaves, and flowers of a pink color in a loose spike, 2 to 4 inches in length, raised on a long, naked peduncle or stalk. The flowers are succeeded by short, single-seeded pods, which are strongly reticulated or marked by raised lines and depressed pits.

This leguminous forage plant has recently been introduced into this country under the name of "aspercet." Esparsette is the German name; sainfoin is the name used in France and England.

It is a native of Central and Southern Europe and Western Asia, and in Europe has long been in cultivation. From experiments made by the Duke of Bedford, in England, we learn that it was first introduced to English farmers as a plant for cultivation from Flanders and France, where it has been long cultivated. It was found to be less productive than the broad-leaved clovers, but on chalky and gravelly soils there was abundant proof of the superiority of sainfoin. It produces but little herbage the first year, but improves in quantity for several years. Mr. Martin J. Sutton, in a recent work on "Permanent and Temporary Pastures," says that it has been cultivated in England for over two hundred years. He says that it is essentially a food for sheep, and in pasturing the sheep do it no injury. It is also useful for horses, but produces nothing like the quantity of green fodder that can be obtained from the Lucerne patch. When sown alone Mr. Sutton says that sainfoin is liable to decrease and become overrun with weeds. He recommends its use as a predominant constituent in a mixture of grasses and clovers. He says that combined with strong growing grasses there is less risk, and the grasses keep down the weeds which otherwise are apt to overrun the sainfoin. In a green state it is quite free from the danger of blowing cattle (hoven), and when made into hay is an admirable and nutritious food. But it requires great care in drying when made into hay.

Mr. Sinclair states that the produce of sainfoin on a clayey loam with a sandy subsoil is greater than on a sandy or gravelly soil resting upon clay.

A French writer says that sainfoin can not accommodate itself to damp soil, which, although dry, rests upon a wet subsoil. It delights in dry soil, somewhat gravelly, and, above all, calcareous. It flourishes upon the declivities of hills where water can not remain, and in light soil, where its powerful roots can readily penetrate. But although surviving in the poorest calcareous soil, like clover and lucerne, its productiveness is always relative to the permeability and fertility of the land. It prefers open, sunny places, with a southern or eastern exposure.

Sainfoin has received several trials in this country, but without much success, probably from the experiments having been made upon unsuitable soil. We can not expect that it will be preferred in places where clover succeeds, but in light soils and in regions with a light rain-fall it should receive a thorough trial. A recent bulletin of the Iowa Agricultural College gives the result of some experiments with this plant which are very satisfactory. Observations there made indicate that it stands early freezing quite as well as Kentucky blue grass. It produces at the rate of 3 tons of dry hay per acre. It deserves trial in Kansas, Nebraska, and Colorado. (Plate 95.)

MEDICAGO.

Medicago sativa (Alfalfa).

This plant is called Lucerne, medick, Spanish trefoil, French clover, Brazilian clover, and Chilian clover. It is not a true clover, though belonging to the same natural family as the clovers. Alfalfa, the name by which it is commonly known in this country, is the Spanish name, which came into use here from the fact that the plant was introduced into cultivation in California from South America under the name of alfalfa, or Brazilian clover. The plant had previously been introduced into the Eastern and Southern States, but attracted little attention until its remarkable success in California. In Europe it is generally known as Lucerne, probably from the canton of Lucerne, in Switzerland, where it was largely cultivated at an early day. It has been known in cultivation from very ancient times, and was introduced from Western Asia into Greece about 500 B. C. It is now largely grown in southern France, and to a considerable extent in other parts of Europe. It has been introduced into several of the countries of South America, and on the pampas of Buenos Ayres it has escaped from cultivation and grows extensively in a wild state. Though known for a long time in the United States, alfalfa is not yet cultivated to the extent that it should be.

In the Southern States east of the Mississippi it is especially desirable that its merits should be better known. The climate of that section is nearly as favorable to its growth as that of southern California, but much of its soil less suitable, hence reports from different localities vary somewhat as to its value.

Climate.—Alfalfa is less hardy than red clover, and is adapted to a milder climate; still, it has stood the winter safely as far north as Vermont, New York, and Michigan, though farther west, where less protected by snow, it winter-kills more or less even as far south as Texas. The young plants are very susceptible to frosts, and the mature plants, if not killed by the cold winters of the Northern States, are so weakened that they endure there for a much shorter period than in milder climates. A cold of 25 degrees is said to kill the tops, but in the Southern States the plant quickly recovers from the effect of frost and grows most of the winter. In the Northern States, even where it endured the winter, the yield is so much less than at the South that it has little or no advantage over the common red clover. Farther south, however, even where both may be grown, alfalfa is often preferred, not only for its larger yield, but also for its perennial character. Alfalfa is especially adapted to dry climates, and withstands drought much better than ordinary clovers.

Soil.—Although alfalfa improves the fertility of the soil, it must have a rich soil to start with, and it therefore is of little value as a renovater of worn-out lands. It prefers sandy soils, if fertile. The failure on sandy soils in the East and the South has been mainly due to the lack of fertility to give the young plants a good start and enable them to become deeply rooted before the advent of drought. On this account it usually thrives best on rich bottom-lands. Lands that are tenacious and hold water are not adapted to its culture unless well drained. Most of the lands in the West upon which it is grown successfully have a permeable subsoil. When the soil permits, its roots penetrate to a great depth. Cases have frequently been observed of their reaching a depth of 12 or 15 feet, and depths of more than 20 feet have been reported. Hence, after the plant is established, the character of the subsoil is of more importance than that of the surface.

Culture.—Sow at any time that the ground is in suitable condition, and when there will be time for the plants to become well established before they are subjected either to drought or extreme cold. In the Northern States the month of May will be about the right time. Farther south, in the latitude of northern Mississippi, September is probably the best month, and in the extreme South, or in the warm valleys of California, any time will answer from fall until spring. The soil should be thoroughly prepared, and the seed sown at the rate of 15 to 20 pounds to the acre. If sown broadcast, about the latter quantity will be required; if in drills, the former amount will be sufficient. If the raising of seed is the main object, 12 or 14 pounds to the acre will give the best results, as the plants will be more vigorous and yield more seed, though they will be coarser and less desirable for feed.

Drill-culture gives the best results, especially if the soil be dry or weedy. The drills may be 12 to 18 inches apart according to the tool to be employed in cultivation. The seed, if sown broadcast, may be sown alone or with grain, but it generally gives the best results when sown alone. It is often sown with oats with good results, but in a wet season it is liable to be smothered out unless the grain is sown quite thin. After the first year the harrow may be employed to advantage, and even a narrow plow, of such form as will not cut the roots too severely, is sometimes used with good effect, especially where the planting is in rows. In all cases where weeds are inclined to appear it is desirable to give some kind of cultivation every year. This is not so important where the plant is irrigated as elsewhere. In much of the country reaching from Texas to the Pacific, irrigation is only essential the first year, or until the roots have penetrated deeply into the soil, though the crop is greatly increased by an abundant supply of moisture at all times. In parts of California and adjoining States alfalfa is grown only by irrigation, and this must sometimes be resorted to, even when not essential for the growth of the crop, in order to kill the gophers, which are liable to destroy the plants by eating off the roots a few inches below the surface. Immediate irrigation will also prevent many of the plants so eaten off from dying.

Alfalfa should be neither mowed nor pastured until it has made a considerable growth and becomes well established.

Harvesting, Feeding, etc.—Alfalfa is perhaps best known in most localities as a soiling plant. For this purpose it has scarcely a superior. It may be cut repeatedly during the season, furnishing a large amount of nutritious forage, which is relished by all kinds of stock. It is said to be less liable than clover to cause slobbers in horses. There is some danger, however, especially to cattle, in feeding it while wet or very succulent, of its causing bloat or hoven. On this account it is a good plan to feed it in the green state in connection with straw or hay, or to let it lie several hours to become partially wilted before being fed.

It is when used as pasture that the greatest danger occurs in the use of alfalfa. Many have used it for years, both for soiling and as pasture, without any injurious results, but numerous instances have been reported where cattle have bloated and died from eating too freely of it when succulent or wet. In some instances cattle have been kept upon it from the time it started in spring until June or July, with no evil results, and then, when the growth has become very rank or been wet with dew or rain, they have been taken with bloat. The danger is greater, as is well known, when cattle are suddenly turned into a rank growth and allowed to eat all they will. If cattle are hungry or have not been accustomed to green food they should not be allowed in such a pasture more than half or three-quarters of an hour. In the dry regions of the West there is less danger in the use of alfalfa for pasture than elsewhere, and it is largely used there for that purpose, especially in the fall after a crop or two of hay has been cut. There is considerable danger, however, of the plant becoming killed out by close or continued pasturing, as it does not stand grazing as well as the ordinary grasses and clovers. For hay, the cutting should be done as soon as the blossoms appear, otherwise it becomes hard and woody. Considerable care is required to cure it properly and prevent the loss of the leaves in drying. The yield is so large and the plant so succulent at the time that it must be cut, that unless there is good weather it is difficult to cure ; on this account it is used less for hay, except in dry climates, than it otherwise would be. The increase in the cultivation of alfalfa has created a good demand for the seed, which has thus become one of the most important items of profit in its cultivation. For cleaning the seeds, F. C. Clark, of Alila, Tulare County, Cal., says :

In this part of the State the ordinary grain-thrasher is used. Some extra screens are used and a few changes made in the arrangement of the cylinder and concave teeth. It is the opinion of some of the experienced alfalfa thrashers that a machine combining the hulling process and some of the machinery of the ordinary thrasher would do better work.

The seed is usually taken from the second crop, and the yield is greater than that from red clover, frequently amounting to 10 or more bushels per acre.

The following reports are given from persons who have grown alfalfa in various parts of the country :

J. R. Page, professor of agriculture, etc., University of Virginia:

I have cultivated alfalfa for forty years, both in the tide-water and Piedmont regions of Virginia, and I regard it as the most valuable forage plant the farmer can cultivate for soiling. It is ready to be mowed by the 1st of May and may be cut three or four times during the season. Grazing kills it out. It should be top-dressed with manure every fall and plastered in the spring and after every mowing.

Thomas S. Stadden, Clarke County, Va.:

Alfalfa is grown here to a limited extent. It does well in favorable localities, but is hard to get set. It lasts four to six years.

H. C. Parrot, Kinston, N. C.:

Alfalfa is adapted to rich, open soils in all the Southern States. It is excellent feed either green or cured. It should be sown in drills 18 inches apart and cultivated the first year. After it is well rooted it will stand drought well and crowd everything else out. It will last from eight to sixteen years, according to soil and location.

J. G. Knapp, United States statistical agent, Limona, southern Florida :

Many persons in Florida have experimented with this plant, so valuable in other regions, but nearly all have failed. Sometimes a plant which has come up in the fall and survived the winter has bloomed, but no roots have lived through the wet, warm months of summer. I remember that in New Mexico, whenever it was desirable to destroy the alfalfa, in order to plow the ground, the surface was covered with water daily for two weeks during the heat of summer. The United States consul at Lambayeque, Peru, states (United States Agricultural Report, 1877, p. 544) that it will not bear water, an abundant irrigation or inundation causing speedy death to the plant. The result in this country has been the same. Alfalfa has invariably perished during the rainy months. All the clovers are affected the same way.

Mr. Knapp incloses a letter from Dr. B. J. Taliaferro, of Maitland, Orange County, the only person in his knowledge who has been successful in growing alfalfa in that region.

Dr. Taliaferro says:

There is no doubt but that alfalfa can be successfully grown in south Florida. My old patch is now twelve months old, and has been cut five times. I am so pleased with it that I have just put in 5 acres more. The great difficulty is getting a good stand. If the ground is not just right the seed will fail. I have failed several times by sowing when the sun was too hot or not hot enough, or when the land was not sufficiently moist. From my short experience I think September is the best month in which to plant. If we plant early in the spring or summer it is almost impossible to keep the crab grass from taking it. I sow in drills 16 or 18 inches apart, and wait for a warm, moist day for sowing. The plant is very delicate at first, and must be kept clean from grass and weeds. I shall try a small piece broadcast this fall, but doubt whether it will prove a success, as crab grass is its greatest enemy in my portion of Florida. The piece I have growing is on high, dry, pine land, such as would be suitable for orange-growing. Alfalfa, having a very long tap-root, would not do on low land. It is very necessary to prepare the land thoroughly. My plan is as follows: After getting the land clean of all stumps, rubbish, etc., I plow it deeply with a two-horse turning-plow, then harrow and hand-rake. Early in spring I put on a light dressing of cotton-seed meal, and sow down in cow peas broadcast, and when the vines are in full bearing I turn them under with a three-horse plow, and as soon thereafter as possible harrow deeply, and broadcast again with some good fertilizer (I prefer cotton-seed meal, bone meal, and potash), harrowing it in well with a spring-tooth harrow. It would be well to repeat the harrowing as often as possible before sowing. About the 1st or middle of September hand-rake perfectly smooth, and put in the seed with a seed-drill, about 6 pounds per acre. Keep clean of weeds and crab grass, and cut when in bloom. A top-dressing of land plaster after the first cutting will prove very beneficial. I have experimented with a number of forage plants, but failed with all except millo maize until I tried alfalfa.

J. S. Newman, Director Experiment Station, Auburn, Ala.:

I have had it fourteen years in profitable growth from one seeding, and have seen it in Gordon County, Ga., twenty-five years old, and still in vigorous and profitable growth. If used for hay it must be cut before it blossoms, or the stems become too woody. Like other leguminous plants it requires especial care in curing, to prevent the loss of its leaves. It may be cut from three to five times in one season, according to the frequency of rains. It is a mistake to suppose that because of its long tap-root it is not seriously affected by drought. It thrives well upon all classes of lands, if fertile and well drained.

Clarke Lewis, Cliftonville, Mass.:

It grows readily in this State on poor, sandy soil, but best on sand loam. It will bear cutting year after year without new seeding, if not too heavily grazed. As a permanent soiling plant it has no superior. It must be cut early, when first coming into blossom; if cut later it becomes woody and makes poor hay. Its introduction has been confined to a few localities.

Prof. James Troop, La Fayette, Ind.:

It is naturalized here, but little cultivated. It is perfectly hardy on our black, sandy loam, but yields no more than Timothy or clover. It will not last here mor than three or four years.

Leonard A. Heil, of the Texas Live Stock Journal, San Antonio, Tex.:

Alfalfa has been successfully raised in this locality only by irrigation, which is practicable to but a limited extent. There are those who claim that it can be successfully grown with only the natural rains, but after careful investigation I seriously doubt its practicability.

James Perry, Whitesborough, northeastern Texas:

Alfalfa is a fair success in our black, waxy soil, and can be cut twice a year, yielding 1 to 3 tons at a cutting. Broadcast sowing is the usual method, and seems to be sufficient on clean land. It stands the drought well and the freeze of ordinary winters. Three years ago, however, I had 7 acres badly killed by "spewing up" in winter, but the scattering plants that remained are doing well.

C. A. Graves, Fiskville, central Texas:

It is cultivated here only to a small extent. It dies out in spots, just as cotton, sweet potatoes, and some other vegetables do, and apparently for the same unknown reason. In some localities, the spots where it dies out cover one-fourth the ground. The uncertainty of moisture on and near the surface for any length of time, owing to hot suns and drying winds, makes the catch from all seeds that germinate near the surface uncertain.

Dr. E. P. Stiles, Austin, Tex., says:

Alfalfa is not permanent here. For two or three years it will produce good crops, and then it begins to die out in circular patches. The spots increase in size until in a year or two they become confluent. Cotton plants sometimes die in the same way, and apple-trees put into such soil are subject to a sudden blight. I have never known alfalfa to be killed by either cold or drought, but its growth is very slight in very dry soil. In Green County it is grown quite successfully under irrigation, but it dies in some localities there the same as here.

J. E. Willett, Farmington, northwestern New Mexico:

Alfalfa grows finely here, and yields so enormously that we want nothing better. We cut it four times during the season, obtaining a ton and a half of hay at each cutting. We raise nothing here except by irrigation. As soon as the crop is taken off, we turn on the water in many places at once and flood the land for several days, for Alfalfa requires an abundance of water, nothwithstanding the fact that land which is low and wet will not answer. It flourishes on rock uplands that are very poor, but must have plenty of water at the right time. The soil is filled with large, long roots, reaching as deep as 20 feet.

George H. Jones, Naranjos, northwestern New Mexico:

It grows well without irrigation after the second or third year on any ordinary soil, and yields very satisfactory results where properly put in. I know one piece which has stood eight years and still yields well.

A. L. Siler, Ranch, Utah:

I know Lucerne patches that have stood for twenty-four years, and they are as productive as when first planted. It does well with irrigation on any porous soil, yielding 4 to 6 tons per acre. Without irrigation it would produce nothing.

William Leaman, Cannonsville, Utah:

Lucerne does very well in this mountain country, where there is very little rain, and produces from 2 to 2½ tons per acre, and makes from three to four crops per year, but I am well satisfied that it will not stand much wet weather, as excessive watering kills it here, and water running over it in the winter and forming ice over it kills it.

Prof. A. E. Blount, Fort Collins, Colo.:

Our soil is mostly sandy loam and clay loam, gray, and to all appearances very poor. It is dry, hard, and destitute of black soil, except in low, marshy places and on the streams. On this soil, which has never been leached or deprived of its fertility by moisture, we sow alfalfa at the rate of 20 pounds to the acre. If kept well irrigated, two crops can be taken the same season that the seed is sown, yielding as high as 3 or 4 tons per acre. The second season, if a good stand was secured, three cuttings are made, yielding as high in some localities as 7 tons. Our largest yields come from those farms where water is applied immediately after each cutting. Among the best farmers 4 tons to the acre is a very small average. I have known 9 tons to be taken from an acre where the most careful attention was given. When once rooted it is next to impossible to eradicate or kill the plant. One man plowed up a piece and sowed it to oats, and after having thrashed out 42 bushels of oats per acre he cut 3 tons of alfalfa hay per acre from the same land. Some have raised wheat, corn, and potatoes with excellent success, after turning under a crop of alfalfa, without in any way interfering with the stand of the latter the next year.

F. W. Sweetser, Winnemucca, Nev.:

Alfalfa is cultivated quite extensively in several parts of the State. It does best in a dark loam. It is hardy and yields, with irrigation, about 5 tons per acre. One season without irrigation will kill it.

O. F. Wright, Temescal, San Bernardino County, Cal.:

Alfalfa is cut from one to six times per year. The yield when good is as follows: First cutting, 2 tons of not very good hay; second cutting, 3 tons of good hay; third cutting, 2½ tons of good hay; fourth cutting, 2½ tons of good hay; fifth cutting, 1 ton of good hay. If the land is v ry dry there may be but one cutting, the roots living, but the tops apparently dead. If it is very dry the roots die also.

Pasturing in the latter part of summer does not injure it much, but in winter and spring, when annual plants are growing, it soon kills it. A good stand can not be obtained without mowing, for worthless weeds would otherwise choke it out. The plants increase in strength for three years.

E. G. Judson, Lugonia, San Bernardino County, Cal.:

Alfalfa is fairly hardy, but it can not stand extreme cold. On dry lands it can not be grown without irrigation. It can be subdued by repeated plowings or keeping away water.

William Schulz, Anaheim, Los Angeles County, Cal.:

Alfalfa fails without irrigation on account of the gophers, which eat off the roots a few inches below the surface. It is one of the best forage plants we have.

William C. Cusick, Union, Oregon:

Alfalfa is not extensively grown in this locality. It is hardy only at the lowest altitudes, or where snow falls deeply. It prefers dry, sandy soils that can be irrigated, on such lands yielding 3 to 4 tons per acre. Without irrigation it is hardly worth cutting. This applies to a portion of the State east of the Cascade Mountains.

A few extracts from various agricultural papers and other publications are here inserted.

Southern Live Stock Journal:

The value of alfalfa in California is inestimable. The plant is eminently adapted to the soil and climate of that State. It is wonderfully productive. It is grown with success in Colorado and some of the Territories, and now and then an isolated report comes up from the great State of Texas that it is fulfilling the highest hopes of those who have given it their attention. Here and there from the Carolinas, Georgia, Florida, Mississippi, Alabama, and Louisiana come favorable reports, but these instances are few and far between. The fact is, alfalfa has never yet had a fair trial in Southern agriculture. Our people, as a people, have never appreciated its value as a worthy addition to southern grasses and forage plants.

The failures that have been made with this plant in the South are doubtless due to the fact that (1) the weeds are allowed to choke it out the first year, or the stock to graze it too closely and bite off the crowns of the plants before the roots were firmly established; (2) the land was not rich enough—it requires very rich land; (3) that the land was not suitable to its growth, or that it held too much water and ought to have been underdrained.

Tulare County (California) Register:

Alfalfa is the foundation of prosperity in Tulare County. It begins to yield the very year it is sown, and increases its yield many years afterward. It will grow where nothing else will, and sends its roots deep down into the moist strata which underlie the top soil all over the country. Alfalfa not only furnishes food for horses, cattle, and sheep, but hogs and poultry thrive upon it as upon nothing else until fattening time comes, when a little Egyptian or Indian corn must be fed to make the flesh solid. In Tulare, alfalfa yields from 6 to 10 tons of hay per acre each summer, besides supplying good pasturage the rest of the season; when it goes to seed it often yields a return of $40 to $60 per acre in seed alone, besides yielding nearly as valuable a hay crop as when not permitted to go to seed. Upon alfalfa and stock, Tulare is building a great and assured prosperity.

George Tyng, in Florida Dispatch:

Sow in any month when the ground is moist and at least four to six weeks before heavy frost or before the season of heat and drought. Less seed will be required if it is soaked before sowing. Put the seed into any convenient vessel and cover with water, not boiling but too hot to be comfortable to the hand, and keep in a warm place for eighteen to twenty-four hours, until the seeds swell enough to partially rupture their dark hulls. When the seeds are ready for sowing drain off all the water through a sieve or bag and dry the seeds with cotton-seed meal, land plaster, or other material, increasing the bulk to a bushel and a half or two bushels for every 20 pounds. If the ground be dry, cultivate just before sowing and sow in the afternoon. Cover as soon as possible, and guard against covering too deeply. The best convenient thing for this purpose is a light drag made of the bushy branches of trees.

Prof. E. W. Hilgard, in the Report of the Department of Agriculture for 1878, page 490, says:

Undoubtedly the most valuable result of the search after forage crops adapted to the California climate is the introduction of the culture of alfalfa, this being the name commonly applied to the variety of Lucerne that was introduced into California from Chili early in her history, differing from the European plants merely in that it has a tendency to taller growth and deeper roots. The latter habit, doubtless acquired in the dry climate of Chili, is of course especially valuable in California, as it enables the plant to stand a drought so protracted as to kill out even more resistant plants than red clover. As a substitute for the latter it is difficult to overestimate the importance of alfalfa to California agriculture, which will be more and more recognized as a regular system of rotation becomes a part of the general practice. At first alfalfa was used almost exclusively for pasture and green-soiling purposes, but during the last three or four years alfalfa hay has become a regular article in the general market, occasional objections to its use being the result of want of practice in curing. On the irrigated lands of Kern, Fresno, and Tulare Counties three and even four cuts of forage, aggregating to something like 12 to 14 tons per acre, have frequently been made. As the most available green forage during the summer, alfalfa has become an invaluable adjunct to all dairy and stock farming wherever the soil can, during the dry season, supply any moisture within 2 or 3 feet of the surface.

Peter Henderson, in an article on alfalfa in the Report of the Department of Agriculture for 1884, page 567, says:

Mr. William Crozier, of Northport, Long Island, one of the best-known farmers and stock-breeders in the vicinity of New York, says he has long considered alfalfa one of the best forage crops. He used it always to feed his milch cows and breeding ewes, particularly in preparing them for exhibition at fairs, where he is known to be a most successful competitor; and he always takes along sufficient alfalfa hay to feed them on while there. Mr. Crozier's system of culture is broadcast, and he uses some 15 or 16 pounds of seed to the acre, but his land is usually clear and in a high state of cultivation, which enables him to adopt the broadcast plan; but on an average land it will be found that the plan of sowing in drills would be the best. Mr. Crozier's crop the second year averages 18 tons, green, to the acre, and about 6 tons when dried as hay. For this section, the latitude of New York, he finds that the best date for sowing is the first week in May; a good cutting can then be had in September. The next season a full crop is obtained when it is cut, if green, three or four times. If to be used for hay it is cut in the condition of ordinary red clover—in blossom; it then makes, after that, two green crops if cut. Sometimes the last one, instead of being cut, is fed on the ground by sheep and cattle.

(Plate 96.)

Medicago denticulata (Bur Clover).

This is a native of the Mediterranean region, which has become naturalized in most warm countries. It was early introduced into California and has become widely distributed in that State, where it is considered of great value.

It is not of first quality either as pasture or hay, but coming at a time of year when other feed is scarce, and often growing where little else will, it is eaten by all kinds of stock. The pods, or burs, are especially sought after in the dry condition, as they remain good until spoiled by rains. Although this plant does not withstand drought as well as many others, it is enabled to grow on dry soils in climates having prolonged drought from its making its growth during the rainy season. Sown early in autumn in the sections to which it is adapted, it grows during the winter and ripens the following spring or early summer. It has been introduced from California into the Southern States, where it is generally highly regarded by those who have tried it, both for grazing and as a renovator of the soil. Being an annual, and ripening early, other crops may be grown on the same land during the summer without interfering with the next growth of the clover. The clover is usually allowed to reseed itself. But little of the seed is sold in the market, and it is usually sown by farmers without being cleared from the burs, or pods. One serious objection to the plant is the liability of the burs to infest the wool of sheep.

There is another species, called spotted medick (*Medicago maculata*), which is often confused with this, and is probably the more common east of the Rocky Mountains, but the two are much alike and of about the same agricultural value.

Only *Medicago denticulata* is mentioned by Professor Watson in his Botany of California as being found in that State.

J. W. Alesworth, Slack Canyon, Monterey County, Cal.:

On the coast, where the climate is moist, bur clover makes a rank growth and is considered good feed late in the season. My place being 40 miles from the coast and 1,410 feet in altitude it only grows here to a limited extent, though it is gradually extending. When I came to this place, in 1870, there was none here. Bur clover is good, rich feed, but is not sought after by stock until the other clovers and alfilaria are gone.

Daniel Griswold, Westminster, Los Angeles County, Cal.:

It is grown in all the lower valleys of California wherever the land is not very salty, but scarcely any is found in the high valleys. It grows large and falls down and curls around so that it is very difficult to mow, but all stock eat it on the ground, green or dry. The seed is never saved, though it is produced abundantly.

O. F. Wright, Temescal, San Bernardino County, Cal.:

It grows here abundantly on high lands, with alfilaria. These are the only plants on such lands that cattle will eat. They are never killed by cold here, but die when dry weather comes. Stock pick on the bur clover while growing (from January to June), and after it dies they hunt for the burs which are on the ground, and in their efforts to get them they roll the old dry stems into rolls often as big as windrows of hay.

S. H. McGinnes, of Belmont, Tex.:

The California bur clover does well here, making good hay and pasture. It comes up in October and ripens in May. It takes but very few bunches to produce a bushel of seeds (burs) and it only has to be planted once. Horses and hogs do well upon the burs after they ripen and fall off.

Edwin C. Reed, Meridian, Miss.:

Bur clover has been grown here to a limited extent, and a few who have grown it twelve or fifteen years find it all that could be desired for winter and spring pasture. All stock eat it freely when they acquire a taste for it, and sheep and hogs eat the burs left on the ground. The plant reseeds itself, but the ground should be plowed and harrowed in August to secure an early winter pasture. It matures the 1st of June, after which peas may be broadcasted on the same land, when it will require no fall plowing. On rich lands it sometimes seeds in Bermuda beds, affording both winter and summer grazing. I have grown vines 6½ feet long, hip high, and as thick as it could stand. I prize it above all other winter pastures. It is admirably adapted to the Eocene formation, where red clover does not succeed, and it is far better if it did, as bur clover is a winter plant.

94

J. S. Newman, Director Experiment Station, Agricultural and Mechanical College, Auburn, Ala. :

First introduced into the cotton States, as far as I know, by the late Bishop George Pierce, from California, about 1867, and planted at his home in Hancock County, Ga. It has since become quite popular in some localities.

(Plate 97.)

DESMODIUM.

Desmodium is a genus belonging to the same family as the pea and clover, and like them is rich in nutritious material. There are about forty species native in the United States, many of them hard and woody, but several of them furnishing valuable woods-pasture to wild and domestic animals. These are often called beggar-tick, beggar-lice, beggar-weed, or tick-weed, from the tendency of the seed-pods to cling to the clothing of persons or the hair of animals. The same or similar names, however, are applied to other plants.

The species of perhaps the most importance is *Desmodium tortuosum*, which is confined to Florida or the vicinity of the Gulf coast. Seeds of this species were distributed by the Department of Agriculture in 1879, under the name of *Desmodium molle*, and a number of favorable reports have been received from those who have tried it in the southern portion of the Gulf States. It is valued most as a renovating crop for lands where clover can not be successfully grown. It is also of considerable value as pasture, and has sometimes been used for hay.

J. G. Knapp, Limona, Fla.:

Few forage plants bear a better reputation here than *Desmodium molle* (*tortuosum*), commonly known as beggar-weed. Horses prefer it to any other growing plant. It comes as a volunteer in fields planted with other crops. When the stalks are 30 or 40 inches high it may be cut for hay, and as many as 2 tons secured from an acre. The stubble will put forth new shoots and mature sufficient seed to restock the field. It will thrive on the poorest sandy soil, and in a few years, if turned under when matured, will render them rich and productive.

J. C. Neal, Archer, Fla.:

It is especially valuable to Florida, as it enriches the soil beyond any other crop and is not in the way of the corn crop, germinating after corn is laid by. Cattle and horses fatten on this plant rapidly; in fact, nothing is better to restore health and vigor to a worn-out beast than a few weeks in a beggar-weed patch. It is of no value for hay or winter forage.

J. A. Stockford, Caryville, Fla. :

It is at home in middle Florida, and is being introduced in western Florida by some enterprising farmers who had a chance to test its value in middle Florida while farming there. Those who have condemned it have usually done so without apparent reason.

D. S. Denmark, Quitman, Brooks County, Ga.:

We have a plant here known as beggar-weed that grows on cultivated lands, and when once seeded always seeds itself. It is a fine summer and fall forage plant; also fine for hay and for renovating worn-out lands, but difficult to exterminate. It grows only in south Georgia and in Florida.

W. B. McDaniel, Faceville, Ga. :

Beggar-tick or beggar-lice grows well in the southwestern part of Georgia, is an excellent plant for forage, both green and cured, and is splendid as a fertilizer, building up land very rapidly. From the 1st of July it will entirely cover the ground the same season.

R. J. Redding, Atlanta, Ga. :

Introduced from Florida and cultivated in southern Georgia for hay and as a renovator of the soil, especially the latter. It is not hardy against cold, and is not grown in middle and northern Georgia.

Whitfield Moore, Woodland, Red River County, Tex. :

That which I cultivated was from seeds from the Department of Agriculture, and appears somewhat different from the native. It has to be seeded annually. It will not stand much grazing, but is a good fertilizer, and drought seems not to affect its growth in the least. It is best adapted to light, sandy land, and will grow a heavy crop from 4 to 6 feet high on the poorest sandy land we have, and in the driest seasons. The hay is very sweet and nutritious, and all stock eat it more greedily than anything else I have ever fed. The only objection to it is the trouble of saving and cleaning the seed.

LESPEDEZA.

Lespedeza striata (Japan Clover).

This plant was introduced in some unknown way, over forty years ago, from China into the South Atlantic States. It was little noticed before the war, but during the war it extended north and west and has since spread rapidly over abandoned fields, along roadsides, and in open woods, and now furnishes thousands of acres of excellent grazing in every one of the Gulf States, and is still spreading northward in Kentucky and Virginia, and westward in Texas, Indian Territory, and Arkansas. It is an annual and furnishes pasture only during summer and until killed by frost in the fall. The small purplish blossoms are produced singly in the axils between the leaf and stem, and the seeds ripen, a few at a time, from about the 1st of August until the close of the season. It reproduces itself from seed on the same ground year after year, and on this account has been erroneously called a perennial. It will grow on poor soils, either sand or clay, but prefers the latter. It is better adapted to poor soils than Bermuda grass, both from giving a more certain and perhaps larger yield, and from being more useful in restoring their fertility. On poor upland soils it is seldom cut for hay, growing only from 6 inches to 1 foot in height, and being inclined to spread out flat upon the surface. On rich bottom-lands it grows thicker, taller, and more upright, and is largely cut for hay. It has been sown artificially only to a limited extent as yet, but seed is now offered in the market, and its cultivation is likely to be liberally extended, especially on lands too dry or poor for alfalfa and where the true clovers do not succeed. Japan clover is remarkable for holding its own against other plants. It will run out broom sedge and other inferior plants, and even Bermuda in some localities. It does not withstand drought as well as either Bermuda or Johnson grass, but soon recovers after a rain. The young plants are easily killed by drought or frost, and for this reason a good catch is more certain on an unbroken sod than on well-prepared land. Still, there is believed to be less difficulty in obtaining a catch with this than with some other forage plants. A good method of seeding is to sow in March at the rate of one-half bushel per acre, on small grain sown the previous autumn or winter.

For hay it should be cut early, before it becomes woody. It is cured in the same manner as clover, and the hay is apparently relished by all kinds of stock. There is some complaint that stock do not at first eat it readily while growing, and that horses and mules are liable to be salivated if allowed to eat it freely while luxuriant. In both these respects, however, it probably differs little from the ordinary clovers. No cases have been reported of bloat or hoven being caused by it.

E. L. Allen, Brownsville, Haywood County, Tenn.:

Lespedeza striata (Japan clover) grows luxuriantly, is very hardy, and is the best pasture we have in summer. It is especially adapted to poor upland, covering the earth, eradicating weeds and sedge grass, preventing land from washing, and increasing its fertility. It grows well in the open timber. Our special need has been a grass to withstand the heats of summer and afford pastures for the early fall. Japan clover has met this requirement.

H. H. Lovelace, Como, Henry County, Tenn.:

Japan clover made its appearance here three or four years ago, and now occupies nearly all lands that have been exhausted and turned out, growing on land too poor to grow any other plant. In fact, it will grow in a red gully; hence it is the best thing to prevent washing I ever saw, besides all kinds of stock are fond of it; and grow fat on it.

B. D. Baugh, State statistical agent, Carrollton, Miss.:

Japan clover is the most wide-spread of the natural forage plants of this State. It grows luxuriantly on any kind of soil except light prairie ash-land. It is easily cured, makes hay of excellent quality, and furnishes more than half of the long forage of this State. It grows well on upland, but best on bottom-land and alluvial soil, where it frequently attains a height of 30 inches. If intended for hay it should be mowed when the first bloom appears, and be "browsed" or stacked after six or eight hours' exposure to the sun. It affords good pasture from the 1st of May until killed by frost, about the middle of November.

George Echols, Longview, Gregg County, northeastern Texas:

It appeared here four years ago, and it now has possession of all the open idle land. It seeds very abundantly, and grows so densely that it forms a mat. It flourishes with Bermuda grass, so that the hay mowed is about half and half.

Dr. D. H. Brodnax, statistical correspondent, Brodnax, Morehouse Parish, La.:

Lespedeza was first noticed here about 1865. It is supposed to have been introduced in the cavalry hay fed the horses of the Federal cavalry, which occupied this parish for a short time. It has since covered nearly the whole parish. It is not cultivated, but is rapidly rooting out nearly every other grass in the parish. It kills our bitterweed (dog fennel), Bermuda grass, and everything else. It is a splendid forage crop, and excellent for grazing until frosts destroy it.

Dr. Charles Mohr, Mobile, Alabama:

Lespedeza striata (Japan clover) is an annual plant, which, during the last twenty years, has spread all over the Gulf States. It blooms and ripens its seeds from the early summer months to the close of the season, and grows spontaneously in exposed, more or less damp, places of a somewhat close, loamy soil. No attempts at its cultivation have been made. In the stronger soil of the lands in the interior this plant, protected from the browsing of cattle, grows from 1½ to 2 feet in height, and yields large crops of sweet, nutritious hay, the same plot affording a cut in August and another in October, yielding, respectively, 1½ tons and 1 ton of hay to the acre. The plant is perfectly hardy, and is not known to have been killed out by a long drought. It is easily subdued by cultivation, as it does not again make its appearance on land where it has been plowed in, and is not found among the weeds the farmer has to contend with in the cultivation of his crop. It is a perfect pasture plant, easily established, and standing browsing and tramping by cattle well. Its propagation through the woods and pastures is effected by cattle, the seeds passing through the animals with their vitality unimpaired. As a fertilizing plant it is greatly inferior to the Mexican clover.

J. B. Wade, Edgewood, De Kalb County, Ga.:

It is said by the old residents here that Japan clover was unknown in this part of the country until "after the war." It now grows spontaneously on most of the land of middle Georgia that has a red-clay subsoil, and which has been turned out, *i. e.*, not plowed or cultivated for two or three years. It grows sufficiently high to make hay, but as it springs up in February, or even earlier should there come a warm spell of weather, it is mostly used for grazing, as it lasts from February to November.

J. B. Darthit, Denver, S. C.:

It does not stand drought as well as Bermuda; both are our best pasture plants. For cattle we have nothing better than Japan clover; but it salivates horses and mules after the 1st of July, especially if very luxuriant.

J. W. Walker, of Franklin, N. C., in a letter to the Blade Farm, says:

Seventeen years ago Japan clover was found here, occupying an area not exceeding 10 feet square. It now covers thousands of acres, upon which all kinds of stock keep fat and sleek, while the yield in milk and beef products has increased a hundred-fold. Our exhausted and turned-out lands that have hitherto yielded nothing but that worse than useless broom sedge (*Andropogon scoparius*), now have in its stead a beautiful carpet of most nutritious verdure.

This plant grows anywhere and on any kind of land, rich or poor, wet or dry, high or low. It has been found in luxuriant growth on the summit of the Blue Ridge, at a height of 4,000 feet. It will catch and grow luxuriantly where none of the clovers proper will grow at all. Unlike them it never runs out.

J. B. McGehee gives the following experiences in a letter to the Southern Live Stock Journal, September, 1886:

This has proved the worst season for its propagation that I have met with. I have this week examined over 200 acres of my last spring's sowing, where I sowed one-half bushel per acre, and I find the most spotted stand I ever saw; and of the whole 200 acres I will get a crop of hay on not to exceed 50 acres. My first sowing of about 80 acres was commenced about March 22, and finished about the 1st of April. This was coming up thickly when the freeze of the 9th of April came, and I am convinced that all seeds then sprouting were frozen out and killed. The sowings during April did better, but anything like a reasonable stand is found only on moist places. The reason for this is the fact that not a drop of rain fell from April 26 to June 6. My worst catch was on comparatively clean land, an oat field, in which the oats had

3594 GR——7

98

been mostly killed by the winter. My best catch was on a grass sod. I found that a freeze or a drought catching the plants before the roots have penetrated the soil are equally disastrous. On some meadows of previous sowings I am now cutting a heavy crop of almost pure Lespedeza. The reverses of this year will not loosen the hold of the grass on my estimation in the least.

(Plate 98.)

MISCELLANEOUS PLANTS.

OPUNTIA.

Opuntia Engelmanni (Nopal; Prickly Pear).

One of the principal characteristics of the vegetation of arid districts is the prevalence of different species of *Cactaceæ* or cactus-like plants. These are exceedingly variable in form and size, and are divided into several genera. Of these the *Opuntias* are extremely common. There are two kinds of these—one with broad, flat joints, and one with cylindrical or club-shaped joints.

Of the flat, broad-jointed kind there are many species. The *Opuntia vulgaris* is common in sandy ground in the Eastern Atlantic States. In western Texas and other parts of the arid regions reaching to California there is a much larger kind, of the same general appearance, which is called *Opuntia Engelmanni*. This is a stout, coarse-looking plant, growing from 4 to 6 feet high, and much branched. The joints are, in large specimens, a foot long and 9 or 10 inches broad, with groups of stout spines from ½ to 1½ inches long. They are apparently leafless, but in young specimens minute, fleshy leaves may be detected. Springing from the side of these joints at the proper season are handsome flowers 2 or 3 inches in diameter, which are succeeded by a roundish fruit, nearly 2 inches long, filled with a purplish pulp, generally of an insipid taste, while imbedded in the pulp are numbers of small, hard seeds. The common name of this *Opuntia* among the Mexicans is "nopal," and some of the species have fruit which is edible and highly esteemed.

The use of the above species of prickly pear, or cactus, for forage in the dry regions of Texas and westward is a matter of considerable importance. An extended account of its use is given in Bulletin 3, of this Division. The usual method of preparing the plant for feeding is to singe the prickles over a brisk blaze. To some extent, especially by sheep, the plant is eaten in the natural state, but serious consequences frequently result in such cases. Its chief use is as a substitute for fodder in times of scarcity, but when properly prepared and fed with hay and grain it forms a valuable article of food for cattle.

J. A. Avent, Sr., Bexar County, southern Texas:

I have been feeding prickly pear for thirty years. It is an excellent feed for cattle if fed with fodder or hay of any kind; when not too full of sap it may be fed alone. If cut in January it can be fed until March 20, but if left standing it is not good feed after the 20th of February. There is nothing that cattle like better than prickly pear when accustomed to it. We feed it only in dry years when grass is scarce. We begin feeding about the 1st of November and continue until the 20th of February.

99

The old stumps with a little corn will fatten cattle very fast. We burn off the thorns in feeding it, but most stock-raisers do not. The apples ripen about the 1st of July and are eaten by almost everything. Hogs get fat enough upon them to render into lard when the crop is good, and it seldom fails.

A. J. Spencer, Uvalde, Tex.:

It is eaten by cattle, sheep, goats, and hogs. They eat it mainly as found on the range, though sometimes the thorns are scorched off. It is considered one of the best native forage plants, especially to carry these stock through the long droughts that occur occasionally in western Texas. It is a partial substitute for water for all stock that eat it. The only injury I have known to result from eating it has been to sheep, and then only when eaten while frozen.

S. S. Jamison, Burnet, Tex.:

It is used extensively in the southwestern part of the State, especially by Mexicans, for wintering work-oxen, cows, and other cattle upon. The thorns are scorched off before feeding, and no harm results from its use unless it be too great a laxness at times. Only one kind is used as far as I know, but it varies in height in different localities. In this country it grows from 6 inches to 2 feet. Farther south it grows taller.

Prof. George W. Curtis, College Station, Tex.:

It is used quite extensively for cattle and sheep. The prickles are singed off, or the whole plant is boiled and fed, mixed with bran. Only the *Opuntia vulgaris*, and perhaps a variety of the same, are used, so far as I know. I have no positive knowledge of any injury to stock from feeding upon it, but from its purgative nature I should be afraid that it might cause abortion in pregnant cows.

Has your attention been called to the use of the prickly-pear cactus as a lubricant? Certain of the Western railroads have used it with excellent results. It is gathered in Texas, shipped to St. Louis, ground up coarsely, and pine tar added to keep the albuminoids from decomposition (I do not know whether anything else is added or not), after which it is barreled and returned. The total cost is 2¼ cents per pound, and it is said to do the work of 6 or 8 cents' worth of grease and rags formerly used. It is especially useful in preventing and cooling hot boxes. If this comes into general use it will open a new field of production.

Leonard A. Heil, San Antonio, Tex.:

The cactus, or prickly pear, grows abundantly in nearly every section of southwest Texas, often reaching a height of 10 or 12 feet. Ever since the settlement of the country by the English, and probably years before, it has been used to supplement grass in times of drought, but now it is being used with other feeds at all times, and especially in the winter. Sheep do well upon it without water, there being sufficient moisture in the leaves. The herder goes along with a short sword and clips the points of the great leaves, so that the sheep can insert its nose, when it readily eats them entire.

Dr. A. E. Carothers, an extensive ranchman of Cotulla, La Salle County, Tex., began feeding prickly pear and cotton-seed meal to four hundred head of steers for the purpose of fattening for the market, and at the last account was highly pleased with the result and confident of financial success. He singes off the thorns with a flame, and cuts up the pear and feeds it mixed, in troughs, with the cotton-seed meal in the proportion of about 5 pounds of meal to 70 pounds of pear. The steers eat this food with great relish and take the food rapidly. They have about a 2,500-acre field to run in. If this method of feeding proves a success, it may work a revolution in this section, as thousands of tons of cotton-seed are exported annually to England, and the supply of the pear is simply inexhaustible. The feeding of the pear need in no way diminish the supply, as whenever a piece of leaf is kept on the ground it takes root and makes another plant, growing rapidly. Corn is always high, and can never be transported here for stock-feed and the stock be shipped back again over the same road with a certainty of success. The utilizing of prickly pear and cotton-seed meal will make beef-raising, as well as breeding, profitable in this portion of the country, and make the ranchmen entirely independent of all other sections.

Dr. Carothers, above mentioned, writes, March, 1887:

In pursuance of a correspondence had with your Department last summer, begun by Mr. A. J. Dull, of Harrisburg, Pa., who has cattle interests in this State, I have fed four hundred beeves, and am now feeding eight hundred more, on this food. From the analysis furnished by Mr. Richardson of your Department, I found that the cactus was deficient in albuminoids, and from the well-known richness of the cottonseed oil cakes in these elements, I selected it to supply the deficiency, which it did very well. At first I burned the thorns off the cactus, then cut it up by a machine which I devised, and spread it in large troughs, scattering the cotton-seed meal over it, when the cattle ate it with great avidity. I soon found, however, that the burning was injurious, as it was impossible to conduct it without cooking the cactus to a greater or less extent, which caused purging in the animals. To remedy this, i. e., to destroy the thorn without scorching, I took advantage of the botanical fact that the thorns of *Opuntia Engelmanni*, the only one I use, are set at an angle of about 60 degrees backward to the plane of the leaf, and that a cut of half an inch would strike every one of them. I therefore set the knives of my machine to a half-inch cut, and find that when cut in this manner cattle eat it fully as well as when scorched, with none of the unpleasant results referred to. I feed per head about 60 pounds of the cactus and an average of about 6 pounds of the meal per day for ninety days. A train-load of three hundred and thirty head of these cattle sold last week in Chicago at 4½ cents per pound. The meat is singularly juicy and tender, the fat well distributed among the muscles. I have sold it at 1 cent per pound gross over grass cattle in San Antonio.

John C. Chesley, Hamilton, Hamilton County, central Texas:

The prickly pear is used here to a great extent. We have a ranch in Stephens County where we are now feeding the pear to over a hundred of our poorest cattle, and they are doing well on it. It is fed at nearly all of the ranches of Stephens County where they are feeding at all, and there are thousands of cattle being fed this winter on prickly pear that are doing well and will come to grass in good shape that would otherwise have died, or at least the larger part of them.

The pear should be cut and hauled to the feed-lots while the sap is in the roots, or before the warm days come, for if it is fed when the sap is in the tops it is liable to cause laxness and weaken the animals. We prepare it for feeding by holding it for a moment over a blaze. I believe that in the southern part of the State they have a burner with which they burn off the prickles without cutting the plants from the ground, and then let the cattle eat them as they please, but we prefer to cut and feed as above stated. One good man can prepare the cactus and feed about a hundred head of cattle in this way. A poor or half-starved animal should be fed only a small quantity of it at first, which may be gradually increased until the animal is allowed to eat all it wants. When fed in this manner to range-cattle we have never known any injurious results. But if it is fed to steers, and they are worked immediately afterwards, even if the feed is small and they are accustomed to it, they are liable to swell up. We have had them do so when we thought there was danger of its proving fatal. They can be given a feed at night, however, and then worked the following morning without danger of any injurious results.

H. J. Hunter, M. D., Palestine, Tex. :

West of the Colorado River, in this State, the cactus grows in vast forests. I have seen cattle and sheep feed on it as it grows wild. Stockmen cut it on the ground, singe off the prickles, and cut in small bits for their stock.

Mr. Alonzo Millett, of Kansas City, Mo. :

I confine the treatment of my stock in La Salle County, Tex., for their first six weeks or two months, in that locality, almost exclusively to the feeding of prickly pear, which simple measure has proved highly successful, and is worthy of more general trial as a preventive of Texas fever. There is a cactus, called by the Mexicans *Nopal de Castilliano*, which is cultivated in this State for its fruit. This plant grows very large and yields enormous crops of fruit, which is sold on the street for food and to make beer. The young growth of the cactus is used in early spring by the Mexicans of western Texas as food. It is cut in small pieces, mixed with flour in a batter, and fried. It is said to be as palatable as egg-plant.

Otanes F. Wright, Temescal, San Bernardino County, Cal.:

Many kinds of cactus grow here. The flat kind, or prickly pear, is abundant in places. Cattle, goats, and sheep eat it sometimes without any preparation when very hungry, but it looks as though needles and pins would be a pleasanter and safer diet. I have never known, however, any bad results to come from eating it. After boiling to soften the thorns it makes good food for milch cows, and is much relished. The trouble of boiling prevents its extensive use.

(Plate 99.)

EUROTIA.

Eurotia lanata (White Sage).

It is a perennial, half shrubby plant, growing a foot or two high, with slender, woolly twigs, which are abundantly covered with linear sessile leaves an inch and a half long, with a velvety surface of a grayish color and with the margin rolled back. They are mostly in small fascicles or clusters. The flowers are minute and in small clusters in the axils of the leaves, chiefly on the upper part of the stem. The flowers are of two kinds, male and female, on separate parts of the stem, or sometimes on separate plants. The small fruit is covered with long and close whitish hairs. The plant belongs to the order *Chenopodiaceœ*, or the same order as the common pig-weed.

The plant known as "white sage," or "winter fat," is abundant in places through the Rocky Mountain region from Mexico to British America. Prof. S. M. Tracy, who visited portions of Nevada, Arizona, and adjoining territory, in 1887, investigating the native forage plants, under the direction of the Commissioner of Agriculture, states that in the more arid districts of Arizona, Nevada, and Utah, this plant, with grease-wood (*Sarcobatus vermiculatus*), are the most highly valued plants for winter forage. An important fact in regard to the plant is its ability to thrive in somewhat alkaline soils. It is employed as a remedy for intermittent fevers. (Plate 100.)

ERODIUM.

Erodium cicutarium (Alfilaria).

This annual, supposed to have been introduced from Europe, does not seem to be mentioned in any work on forage plants. It occurs abundantly and is of much value for pasture over a large extent of territory in northern California and adjoining regions; elsewhere in the United States it is sparingly introduced and usually regarded only as a weed, though it is not very troublesome. Besides the above name it is known as storksbill, pin clover, pin grass, and filaree; it is neither a grass nor a clover, but belongs to the geranium family; it starts very early, grows rapidly, furnishing good early pasture, and ripens seed before the hottest weather; it is of little value as hay, and is not worth introducing where the ordinary forage plants can be grown. The seed is seldom sown, but the plant comes spontaneously each year from self-sown seed. A few have begun its artificial propagation, and it is undoubtedly worthy of introduction into other regions in the South and West having prolonged droughts; it is hardy at the North, but makes a much smaller growth there.

Brewer and Watson, in "The Botany of California" say in regard to it:

Very common throughout the State, extending to British Columbia, New Mexico, and Mexico; also widely distributed in South America and the Eastern Continent. It has generally been considered an introduced species, but it is more decidedly and widely at home throughout the interior than any other introduced plant, and according to much testimony it was as common throughout California early in the present century as now. It is popularly known as *alfilaria*, or less commonly as pin clover and pin grass, and is a valuable and nutritious forage plant, reputed to impart an excellent flavor to milk and butter.

Prof. E. W. Hilgard, in an article on the Agriculture and Soils of California, in the Report of the Department of Agriculture for 1878, page 488, says:

Two species of crane's-bill (*Erodium cicutarium* and *moschatum*) are even more common here than in Southern Europe, and the first-named is esteemed as one of the most important natural pasture plants, being about the only green thing available to stock throughout the dry season, and eagerly cropped by them at all times. Its Spanish name of *alfilerilla* (signifying a pin, and now frequently translated into "pin weed") shows that it is an old citizen, even if possibly a naturalized one.

Otanes F. Wright, Temescal, San Bernardino County, Cal.:

Alfilaria grows plentifully and is native here. It is the best grass that we have during the wet season while green, but does not amount to much when dry, for it shrinks much in drying, and when dry breaks easily into very fine bits, almost to dust.

Alfilaria and bur clover nearly always grow together on the same land; cold weather never kills either of them. Stock pick for the alfilaria while growing (from January to June), but after it dies they hunt for the clover-burs which are on the ground, and in their efforts to get the burs they roll the old dry stems into rolls, sometimes as big as windrows of hay.

Bur clover and filaria (alfilaria) grow on high land, and die when dry weather comes.- I do not know but they might be kept green all the year if kept wet.

They are about the only plants which grow on the high land all the year as alfalfa does on the low lands. As nine-tenths of our land is dry land, you can see the extent of our needs.

Daniel Griswold, Westminster, Los Angeles, Cal.:

I think alfilaria would be a good thing to raise in the Southern States, but it will be a rather hard seed to gather, though not so hard as Bermuda grass. It produces a small-jointed seed, with a beard or curl attached. Butte or Colusa County would, be the best place to obtain the seed. The plant is native here. It is never cultivated but comes up of itself whenever there is rain enough. It grows everywhere (except in swamps) in damp land, on the driest land, and on the tops of hills up to the snow-line. It has a root that runs slightly downward, and it has to be very dry to prevent it making seed. On damp, rich land it grows large enough to make a good swath of hay. On poor or dry land it is small and dries up. but even in its dry state stock eat it clean and are very fond of it.

C. R. Orcutt, San Diego, Cal.:

Erodium cicutarium and *Erodium moschatum* (about equally used) grow abundantly in southern California and through northern lower California, sometimes attaining a height of 2 feet or more. They grow on dry lands, but only in wet years or where there is abundant rain-fall do they attain any size.

O. F. Thorton, Phœnix, Maricopa County, Ariz.:

It is not cultivated, but is rapidly spreading on the dry ranges (*i. e.*, valleys and mountain sides), and is one of the very best wild grasses, either green or dry.

(Plate 101.)

RICHARDSONIA.

Richardsonia scabra (Mexican Clover; Spanish Clover; Florida Clover; Water Parsley; Bell-fountain; Poor Toe; Pigeon-Weed, etc.).

This is an annual plant of the family *Rubiaceæ* which contains the coffee, cinchona, and madder. It is therefore not a true clover, that name having perhaps been given from the general appearance of the plant and the fact that the flowers are mostly borne in terminal heads. The stem is spreading, branching, and somewhat hairy, and the leaves, unlike the clovers, are composed of a single piece. The plant is a native of Mexico and South America, which has become naturalized in the United States, especially along the Gulf coast, where its chief value seems to be as a renovator of poor, sandy soils. In more dry, exposed regions it seems to require rich, cultivated soils in order to do well. It has been but little cultivated, and it is not known how far north it may be grown successfully, but it would probably have little value where clover can be readily grown. The statements in regard to its value for pasture and hay are very conflicting. It is usually quite succulent and not readily cured in the climate where it is most largely grown. As it grows chiefly in cultivated grounds, it is often looked upon only as a weed.

B. E. Van Buren, Lakeside, Fla.:

I have disseminated the Spanish clover all over my place, as I consider it a valuable plant for improving the land. It is also a very good forage plant, and will grow on the poorest soil without manure.

J. C. Neal, M. D., Archer, Fla.:

Grows rapidly, seeds itself, and makes a fair looking lawn or field, but I have not found a cow or horse that would touch it green or dry.

J. G. Knapp, Hillsborough County, southern Florida.:

Found in moist fields in this county and considered a valueless weed. It is not eaten green by either cattle or horses, and grows flat on the ground, so that it can not be cut for hay. On account of the large number of seeds it perfects it is difficult to eradicate. It is spoken of in some sections as a fertilizing plant. In my opinion it has no other value, and I estimate it low for that purpose.

B. C. Smith, Cold Water, Ga.:

Thrives only on highly fertilized soils, in the best of tilth, where it gives a large yield. Mexican clover, being very similar to purslane, is very hard to cure, and is not well relished by cattle or horses.

C. Menelas, Savannah, Ga.:

I have seen it only on the Gulf coast, where it flourishes luxuriantly without cultivation, and is dreaded by nearly every one as a weed. Stock appear to be very fond of it, and the yield per acre must be very heavy.

Dr. Charles Mohr, Mobile, Ala.:

Introduced from the neighboring tropics and perfectly naturalized. It is never cultivated, but takes possession of the fields, and arrives at the period of its fullest growth after the crops of vegetables, Irish potatoes, corn, and oats, are laid by or have been removed, yielding spontaneous crops of hay and affording fully two cuttings during the season of from 1 to 2 tons per acre, according to the fertility of the field.

In 1874 the same gentleman sent a sample of hay of this plant to the Department, which was found to be nearly as rich in food elements as clover hay. In his letter he then said:

It forms a large and important part of the pine-woods pasture in this county. It is much relished by horses and mules, which seem to thrive well upon it, and sheep feed upon it with great avidity. The plant is known here by the name of "Mexican clover," "poor toes," or "pigeon-weed." Seventeen years ago it was but sparse; now it occurs in all our cultivated grounds, covering them with a luxuriant vegetation after the crops of the summer have been removed.

Thomas J. Key, editor Southern Agriculturist, Montgomery, Ala.:

It grows luxuriantly on cultivated, sandy lands in the southern part of the State, makes excellent hay, and matures after corn has been laid by.

James B. Siger, Handsborough, southern Mississippi:

Of late years Mexican clover has been introduced and grown among the crab grass. It is spreading rapidly. Its habits and manner of cultivation are the same as crab grass. Cattle will pick it out from any other hay and eat it in preference to any.

Edward C. Reid, Meridian, Miss.:

It is hardy, and grows on the poorest sandy land from the coast up to the Cretaceous formation. It stands drought and is hard to exterminate. It comes up after corn is laid by, and on cotton-land covers the cotton. It is not especially valuable as a pasture plant, as it comes up late and pasturing kills it out. In cultivated lands it reseeds itself, and comes up year after year.

Clarke Lewis, Cliftonville, Noxubee County, Miss.:

It grows in the Gulf States, on sandy land, and furnishes abundant forage of fair quality on poor soil. There is none in this section.

W. H. Nevill, Binnsville, Miss.:

Does well in the southern half of the Gulf States.

J. H. Murdock, Bryan, Brazos County, central Texas:

It is grown here and stands drought very well on our light, sandy soils, and makes good pastures in its season.

Mr. Matt. Coleman, Leesburgh, Sumter County, Fla., in 1878, wrote to the Department:

The tradition is, that when the Spanish evacuated Pensacola this plant was discovered there by the cavalry horses feeding upon it eagerly. Five years ago I procured some of the seed and have since grown it in my orange groves as a forage plant and fertilizer. It grows on thin pine land 4 to 6 feet in length, branching, and forming a thick mat, which affords all the mulch my trees require. It requires two days' sun to dry it, and its sweet hay is relished by horses and cattle. The white bloom opens in the morning and closes at evening, and is visited by bees and butterflies.

(Plate 102.)

So much interest is now felt in the matter of new varieties of grasses, especially by the Western experiment stations, that it is thought best to add descriptions and figures of some additional species which have been recommended for trial.

The Colorado experiment station, aided by this Department, gave especial attention last summer to the collection of seeds of the native grasses of that region, and some fifty kinds were selected, and will be subjected to cultivation on the arid land of that section.

Other western stations will take up the same line of work in the future. The illustrations given are especially valuable for the identification of the various species by students or by any persons who are interested in the subject.

Panicum gibbum.

A perennial species, growing in swamps and low, wet ground in the Southern States from North Carolina to Florida. The stem is decumbent, branching and rooting at the lower joints. The panicle is 3 to 5 inches long, and narrow, the branches being appressed. The leaves are smooth or smoothish, half an inch broad, and 6 to 8 inches long. The whole grass is of a deep green color. The flowers drop off soon after flowering. The grass, if it occurs in abundance, would be of considerable value, as it furnishes a good deal of nutritious matter.

Mr. J. H. Simpson, of Manatee, Fla., writes as follows:

This most valuable grass seems to have been entirely overlooked as far as its qualities for hay and pasturage are concerned. It is perfectly at home in any situation. It usually grows in wet places, with culms 2 or 3 feet high. The late J. N. Harris informed me that he believed that from 3 to 5 tons of most excellent hay could be cut per acre, and that it was an excellent pasture grass. He had experimented with it for years.

(Plate 103.)

Muhlenbergia comata.

This species is closely related to *M. glomerata*. It grows throughout the Rocky Mountain region in Colorado, Wyoming, Utah, Idaho, and California, usually on the sandy and alluvial banks of streams. It grows in tufts from firm, creeping root stocks. The culms are erect, 2 to 3 feet high, and leafy below. The panicle is 2 to 4 inches long, narrow and close, sometimes interrupted below, generally of a dark lead-color, and of soft texture. The outer glumes are very narrow and acute, and the flowering glume is surrounded at the base by a copious tuft of silky hairs. The slender awn of the flowering glume is three or four times its length.

(Plate 104.)

Sporobolus heterolepis (Bunch Grass; Wire Grass).

This is called bunch grass and wire grass from the abundant, long, wiry leaves and stems. I found it a considerable element in the prairies of southern Dakota, and it occurs southward to Texas. It was also common on the prairies of Illinois and Wisconsin before the incoming of settlements. West of the 100th meridian, however, especially in sandy soils, this species is replaced by two others of the same genus, viz: *Sporobolus cryptandrus* and *S. airoides.* All these species should receive attention.

It grows in dense, firmly rooted tufts, principally west of the Mississippi River, from British America to Arkansas. The panicle is from 3 to 6 inches long, rather loose; the branches, two to three together, slender, and with a few rather distant flowers.

A writer in the Agricultural Report for 1870 says:

This species may be identified from its long, slender leaves, growing abundantly from the base of the plant, gracefully curving; from its tendency to grow in bunches or stools, and when in fruit from its small panicle of sharp-pointed spikelets and its round seeds. These, when bruised, emit a strong, heavy, and rather disagreeable odor. It is sometimes cultivated for hay, and makes an article of fine quality.

Sporobolus airoides (Bunch Grass; Salt Grass).

Culms (arising from strong perennial creeping root-stocks) 2 to 3 feet high, thickened at the base and clothed with numerous long, rigid, generally involute, long-pointed, smooth leaves, which are bearded in the throat of the sheath. The panicle is 6 to 12 inches long and 3 to 4 inches wide, thin and spreading; the branches capillary, and scattered or in whorls below, subdivided above the middle, and rather sparsely flowered.

It is common on the arid plains of the West, is sometimes called salt grass, and affords persistent pasturage where other grasses are tramped out. (Plate 105).

Agrostis exarata, var. **Pacifica** (Pacific Coast Redtop).

This variety grows chiefly on the Pacific coast, from California to Alaska. It is often more robust than the common or eastern redtop, growing 2 to 3 feet high, with a stout, firm culm, clothed with three or four broadish leaves 4 to 6 inches long. The panicle is 4 to 6 inches long, rather loose, heavier, and closer than the proper species.

There is reason to believe that this species can be made to supply the same valuable place on the Pacific coast that the *A. vulgaris* does at the East. It deserves trial. (Plate 106).

Deschampsia cæspitosa (Hair Grass).

This is an exceedingly varied species, having a wide distribution in this and other countries. It is somewhat rare east of the Mississippi, but on the elevated plains of the Rocky Mountains and in California and Oregon it is one of the common bunch grasses which afford pasturage to cattle and horses. At the East is is found in the hilly regions of New England and the Alleghanies. It grows in bunches, which are firmly rooted. The culms are 2 to 4 feet high. The root-leaves are very numerous, long, and narrow. The panicle is very handsome, presenting a purple and glossy hue, and a loose, graceful appearance. Its culms are too light for hay-making, but the abundant root leaves may make it valuable for pasturage, especially in the arid districts. (Plate 107.)

Chloris alba.

An annual grass, growing in tufts, 2 to 2½ feet high, smooth, the culms branching and bent at the lower joints; the leaves are numerous and rather broad, the upper sheaths dilated and at first inclosing the flower spikes, which are in a close cluster, eight to fifteen in number and 2 to 3 inches long. The flowers are sessile and crowded in two rows on one side of the spikes. It is a common grass in the arid districts of Texas, New Mexico, and Arizona. It furnishes a large amount of foliage, and may prove useful in localities to which it is adapted. (Plate 108.)

Diplachne dubia.

A perennial grass of vigorous growth, growing 3 or 4 feet high, the culms rather stout and erect, with an abundance of foliage, the leaves being quite long and narrow. The panicle is from 6 to 12 inches long, consisting of from 10 to 20 narrow, spreading spikes, each 4 to 6 inches long, mostly scattered on the axis, or two or three together. The spikelets are three to five-flowered, the empty glumes linear-lanceolate and acute. The flowering glumes are oblong, obtuse, two-lobed, and smooth except on the margins.

Its principal range is in the Southwest, from Texas to Arizona. It is a promising grass, and should receive the attention of agriculturists. (Plate 109.)

Melica.

Spikelets two to many-flowered; the flowers usually convolute around each other, the upper one small and imperfect; the empty glumes are membranaceous and awnless, the lower one three to five-nerved, the upper five to nine-nerved, the lateral nerves not reaching to the apex. The flowering glumes are of thicker texture, becoming coriaceous, scarious near the apex, mostly rounded on the back, five to nine-nerved, the lateral nerves not reaching the apex, the central one sometimes ending in a short point or even in a long awn; the palets shorter than their glumes, two-keeled and ciliate on the keels. Of this genus we have ten or twelve species.

Melica diffusa.

A perennial species, growing in rocky woods or ravines throughout the Rocky Mountains in Colorado and New Mexico. It grows in loose tufts, the culms about 2 feet high, the lower leaves and sheaths soft hairy, the upper leaves narrow, 3 to 4 inches long and pointed. The panicle is 6 or 8 inches long, open, with rather few (6-8) branches, 3 to 4 inches long, rather distant from each other, and somewhat spreading; the spikelets are large, 4 to 6 lines long, and three to five-flowered, the upper flower imperfect. The empty glumes are quite unequal and much shorter than the spikelets. The flowering glumes are many-nerved below, with a broad scarious margin above. The palets are narrower and shorter than the flowering glumes and fringed on the keel.

This grass is relished by cattle, but as its preference is for shaded places it may not be adapted for general culture. (Plate 110.)

Melica bulbosa.

This species is distinguished by its large bulbous roots, or, more properly, by the bulb-like enlargement of the base of the stem. It grows 2 to 3 feet high, the leaves narrow, scabrous, and becoming involute. The panicle is narrow, from 4 to 6 inches long, with short appressed branches. The spikelets are about half an inch long, with five to seven perfect flowers; the empty glumes are three to four lines long, or nearly as long as the flowering glumes, which are oblong-lanceolate, seven-nerved, and obtuse or notched at the apex.

This species grows in Oregon, Washington, California, Nevada, Montana, Utah, and Colorado. (Plate 111.)

Melica imperfecta.

This is one of the commonest grasses throughout California, particularly in the southern portion. There are several varieties, which differ considerably in size and general appearance. The culms are from 1 to 3 feet high, rather slender and wiry, the leaves rather numerous and narrow. The panicle varies from 6 to 12 inches in length, sometimes narrow, but usually spreading, with rather long and distant branches, which are whorled about in threes of unequal length and numerously flowered. The spikelets have one or two perfect flowers, with a short, club-shaped, imperfect one. The empty glumes are shorter than the flowers and three to five-nerved, the flowering glume about seven-nerved, usually purplish and acute. It is not bulbiferous.

The larger forms of this grass certainly deserve trial for culture in California. (Plate 112.'

Uniola.

This genus has closely many-flowered spikelets, usually large, very flat, and two-edged, one or more of the lower flowers consisting only of an empty glume; the empty glumes are closely folded together, keeled, rigid, or coriaceous; the flowering ones are of similar texture, but larger, many-nerved, usually acute and entire, the palet rigid, with the keels narrowly winged.

Uniola latifolia (Broad Leaved Uniola; Large-flowered Oat Grass).

This is a handsome grass, growing 2 to 3 feet high, with very broad leaves and a large spreading panicle. The drooping spikelets are larger than those of any other North American grass, being an inch or more long and half as wide, consisting of ten or twelve flowers.

It grows from Pennsylvania to Kansas and southward.

Dr. Charles Mohr, Mobile, Ala., says of this grass:

A fine vernal grass with a rich foliage, blooming early in May; frequently in damp, sandy loam, forming large tufts. This perennial grass is certainly valuable, affording an abundant range early in the season; if cultivated it would yield large crops ready for cutting from the 1st of May. It is called by some wild fescue or oat grass. I am not able to judge of its value as a pasture grass.

(Plate 113.)

Glyceria Canadensis (Rattlesnake Grass; Tall Quaking Grass).

Of this genus two species have been already described. This species belongs to the northern portion of the United States, usually found in mountainous districts, in swamps, and river-borders, growing in large tufts. The culms are stout, about 3 feet high, smooth and leafy. The leaves are 6 to 9 inches long, the lower ones often twice as long and quite broad. The panicle is large and diffuse, 6 to 9 inches long, the branches long, slender, and drooping. The branches are more or less whorled, mostly in threes, the largest 3 to 4 inches long, and often subdivided. The spikelets are oblong or ovate, rather turgid, usually six to eight-flowered.

This is quite an ornamental grass. Cattle are fond of it, both green and when made into hay. It is well adapted to low meadows and yields a large quantity of foliage.

Hon. J. S. Gould says:

It is usually found at high elevations, in swampy land, and by the margin of streams. It is very apt to grow in clumps. It is one of the most beautiful of grasses, and is exceedingly ornamental in grass boquets. It is abundant on the Catskill and White Mountains, and on the Raquette waters of the Adirondacks. Cattle eat it very well in pasture and when made into hay.

(Plate 114.)

INDEX.

	Page.
Agropyrum glaucnm...	7, 75
Agropyrum repens.	7, 75
Agropyrum teneram	76
Agrostis alba.	46
Agrostis canina	47
Agrostis exarata	47
Agrostis exarata var. Pacifica	107
Agrostis stolonifera	46
Agrostis vulgaris	46
Alfalfa.	84
Alfilaria	102
Alkaline grass	61
Alopecurus geniculatus	40
Alopecurus geniculatus, var. aristulatus	40
Alopecurus occidentalis	41
Alopecurus pratensis	40
Ammophila arundinacea	43
Andropogon furcatus	7, 35
Andropogon Hallii	35
Andropogon macrourus	35
Andropogan provincialis	35
Andropogon scoparius	7, 35
Andropogon Virginicus	35
Anthoxanthum odoratum	40
Aristida purpurea	41
Arrhenatherum avenaceum	52
Arrow grass	42
Arundo donax	60
Arundinaria tecta	78
Austin grass	25
Avena fatua	52
Barley, wild	76
Barnyard grass	27
Beach grass	48
Beckmannia crucaeformis	24
Bell-fountain	103
Bent grass	46
Bent grass, Rhode Island	47
Bermuda grass	54
Blanket grass	23
Blue grass, Kentucky	66
Blue grass, Texas	64
Blue grass, English	65
Bluejoint	48, 75
Blue stem	35, 75
Bouteloua oligostachya	7, 57
Bouteloua racemosa	7, 57
Brizopyrum spicatum	62
Bromus ciliatus	74
Bromus Schraderi	74
Bromus secalinus	73
Bromus unioioides	73
Bromus Willdenovii	74
Broom corn	37

	Page.
Broom grass, heavy-topped	35
Broom sedge	35
Buchloë dactyloides	7, 59
Buffalo grass	25, 57, 59
Bunch grass	6, 35, 42, 72, 107
Bur clover	92
Cactus	98
Calamagrostis Aleutica	7
Calamagrostis Canadensis	7, 48
Calamagrostis longifolia	7, 49
Calamagrostis neglecta	7
Calamagrostis sylvatica	7, 49
Canary grass, reed	38
Canary grass, southern reed	39
Canary grass, Stewart's	39
Cane, small	78
Cane, switch	78
Ceratochloä unioloides	71
Cheat	73
Chess	73
Chloris alba	108
Chrysopogon nutans	7, 36
Cinna arundinacea	47
Cinna pendula	47
Clover, Alsike	81
Clover, Brazilian	81
Clover, bur	92
Clover, Chilian	81
Clover, Dutch	82
Clover, Florida	103
Clover, French	82, 84
Clover, Japan	95
Clover, mammoth	80
Clover, Mexican	103
Clover, pin	103
Clover, red	79
Clover, running buffalo	83
Clover, Spanish	103
Clover, white	82
Cord grass	56
Corn, broom,	37
Couch grass	75
Crab grass	27, 58
Crowfoot	62
Cuba grass	36
Cut grass	34
Cynodon Dactylon	54
Dactylis glomerata	62
Dactyloctenium Ægyptiacum	58
Darnel, poison	75
Deschampsia caespitosa	6, 107
Desmodium molle	94
Desmodium tortuosum	94
Devil's knitting needles	42

	Page.		Page.
Deyeuxia Canadensis	48	Lespedeza striata	05
Deyeuxia sylvatica	49	Lolium perenne	74
Diplachne dubia	108	Lolium temulentum	75
Distichlis maritima	7, 61	Louisiana grass	23
Dourra	37	Lucerne	81
Egyptian grass	36	Lyme grass	77
Eleusine Ægyptiaca	58	Manna grass, floating	70
Eleusine Indica	58	Marsh grass	56
Elymus Canadensis	77	Meadow grass, fowl	67
Elymus condensatus	77	Meadow grass, nerved	70
Elymus triticoides	77	Meadow grass, reed	69
Elymus Virginicus	77	Meadow grass, rough-stalked	68
Eragrostis Abyssinica	61	Meadow grass, tall	69
Eragrostis major	61	Meadow oat grass	52
Erodium cicutarium	102	Mean's grass	36
Esparsette	83	Medicago denticulata	92
Euchlæna luxurians	30	Medicago maculata	93
Eurotia lanata	101	Medicago sativa	84
Evergreen grass	52	Medick	84
False rice	34	Melica bulbosa	108
Fescue, meadow	71	Melica diffusa	108
Fescue, sheep's	72	Melica imperfecta	109
Fescue, tall	71	Mesquite	6, 57
Festuca elatior	71	Milium effusum	43
Festuca ovina	7, 72	Millet, Arabian evergreen	36
Festuca pratensis	71	Millet, Australian	36
Festuca scabrella	7, 72	Millet, cat-tail	30
Festuca unioloides	74	Millet, Egyptian	30
Filaree	102	Millet, evergreen	36
Finetop	46	Millet, German	29
Florin	46	Millet, Morocco	36
Fowl meadow grass	67	Millet, pearl	30
Foxtail, meadow	40	Millet, Texas	25
Foxtail, Rocky Mountain	41	Millet grass	27
Foxtail, water	40	Millo maize	37
Gama grass	30	Millo maize, red	38
Gietta grass	34	Muhlenbergia comata	106
Gilbert's relief grass	39	Muhlenbergia diffusa	43
Glyceria arundinacea	69	Muhlenbergia glomerata	43
Glyceria Canadensis	109	Muhlenbergia Mexicana	43
Glyceria fluitans	70	Muhlenbergia, spiked	43
Glyceria nervata	70	Muhlenbergia sylvatica	44
Goose grass	25	Munroa squarrosa	7
Grama, black	34	Nimble Will	43
Grama grass	6, 57	Oat grass, large-flowered	100
Green Valley grass	36	Oat grass, meadow	52
Guinea grass, Alabama	36	Oat grass, tall	52
Guinea grass, true	25	Oats, wild	36, 52
Hair grass	107	Onobrychis sativa	83
Herd's grass (of New England and New York)	44	Opuntia Engelmanni	98
Herd's grass (of Pennsylvania)	46	Opuntia vulgaris	98
Hilaria Jamesii	34	Orchard grass	63
Holcus lanatus	50	Oryzopsis cuspidata	6, 42
Hordeum jubatum	76	Panic grass, two-edged	28
Hordeum murinum	76	Panic grass, tall	28
Hordeum pratense	77	Panicum agrostoides	28
Hungarian grass	29	Panicum anceps	28
Indian rice	33	Panicum barbinode	26
Johnson grass	36	Panicum Crus-galli	27
June grass	66	Panicum gibbum	106
Kœleria cristata	6, 66	Panicum jumentorum	25
Leersia hexandra	34	Panicum maximum	25
Leersia oryzoides	34	Panicum milliaceum	27
Leersia Virginica	34	Panicum proliferum, var. geniculatum	26
		Panicum, redtop	28

	Page
Panicum sanguinale	27
Panicum Texanum	25
Panicum virgatum	28
Para grass	26
Parsley, water	103
Paspalum dilatatum	22
Paspalum distichum	24
Paspalum, hairy-flowered	22
Paspalum ovatum	22
Paspalum platycaule	23
Penicillaria spicata	30
Pennisetum spicatum	30
Phalaris arundinacea	38
Phalaris intermedia	39
Phalaris intermedia, var. angusta	39
Phleum pratense	44
Pigeon grass	30
Pigeon weed	103
Pin clover	102
Pin grass	102
Pragmites communis	60
Poa andina	69
Poa arachnifera	64
Poa compressa	65
Poa pratensis	66
Poa serotina	67
Poa tenuifolia	68
Poa trivialis	68
Poa trivialis, var. occidentalis	69
Poor toe	103
Porcupine grass	42
Prickly pear	98
Quack grass	75
Quaking grass, tall	109
Randall grass	71
Rattlesnake grass	109
Red millo maize	38
Redtop	46
Redtop, Pacific coast	1 7
Redtop Panicum	28
Reed grass	50
Reed grass, giant	60
Reed grass, small	48
Reed grass, wood	47
Relief grass, Gilbert's	39
Rescue grass	73
Rice, false	34
Rice, Indian	33
Rice, wild	33
Richardsonia scabra	103
River grass	25
Rye grass	77
Rye grass, giant	77
Rye grass, Italian	74
Rye grass, wild	77
Rye, wild	77
Saccatoo	42
Sago, white	101
Sainfoin	83
Salt grass	56, 61, 107
Sand grass	48
Sarcobatus vermiculatus	101
Schrador's grass	73
Setaria glauca	30

	Page.
Setaria Italica	29
Setaria viridis	30
Slough grass	24
Smut grass	45
Soft grass	50
Sorghum halepense	36
Sorghum, sugar	37
Sorghum vulgare	37
Spartina cynosuroides	56
Spartina juncea	56
Spear grass	66
Sporobolus airoides	107
Sporobolus cryptandrus	45
Sporobolus heterolepis	107
Sporobolus Indicus	45
Sprouting crab grass	26
Squirrel-tail grass	76
Stipa avenacea	41
Stipa comata	6, 41
Stipa pennata	42
Stipa setigera	6
Stipa spartea	6, 41, 42
Stipa viridula	6, 41
Storksbill	102
Sweet vernal grass	39
Switch cane	78
Switch grass	28
Teff	61
Teosinte	31
Terrell grass	77
Timothy	44
Timothy, California	30
Trefoil, Spanish	84
Trifolium hybridum	81
Trifolium incarnatum	82
Trifolium medium	80
Trifolium pratense	79
Trifolium repens	82
Trifolium stoloniferum	82
Triodia acuminata	59
Triodia seslcroides	59
Triodia stricta	59
Triodia Texana	59
Tripsacum dactyloides	30
Trisetum palustre	51
Trisetum subspicatum	51
Uniola, broad-leaved	109
Uniola latifolia	109
Velvet grass	50
Velvet mesquite	50
Vernal grass, sweet	40
Water parsley	103
White grass	34
White grass, small-flowered	34
White sago	101
Wild barley	76
Wild oats	36, 52
Wild rice	33
Wild rye	77
Wire grass	35, 107
Wood reed grass	47, 58, 65
Yard grass	58
Zacate	34
Zizania aquatica	33

NOTE.

It has been thought well to attach to the foregoing, in the form of an appendix, a report of Mr. Clifford Richardson, formerly Assistant Chemist of the Department of Agriculture, on the Chemical Composition of American Grasses, from investigations made by him in the laboratory of the Department of Agriculture, 1878–1882.

The appendix also includes, for the benefit of those readers who may wish to familiarize themselves with them, a glossary of the botanical terms used in describing grasses.

APPENDIX.

THE CHEMICAL COMPOSITION

OF

AMERICAN GRASSES

FROM

INVESTIGATIONS IN THE LABORATORY OF THE DEPARTMENT OF
AGRICULTURE, 1878-1882.

BY

CLIFFORD RICHARDSON,
FORMERLY ASSISTANT CHEMIST.

THE CHEMICAL COMPOSITION OF AMERICAN GRASSES.

In submitting grasses to chemical analysis, with a view of judging of their nutritive value, it is usual to determine the amount present of water, ash, fat or oil, fiber, and nitrogen. From the latter the amount of albuminoids to which it is equivalent is readily calculated by multiplying by a factor which represents the per cent. of nitrogen present in the average albuminoid, and by substracting the sum of all these constituents from one hundred, the percentage of undetermined matter is obtained, and as it of course contains no nitrogen, and consists of the extractive principles of the plant, it is described as "Nitrogen free extract." It includes all the carbo-hydrates, such as sugar, starch, and gum, together with certain other allied substances, with which we are less intimately acquainted, but which have a certain nutritive value.

Although it has been customary to state as albuminoids the equivalent of the nitrogen found, this is rarely entirely correct, as a portion is generally present in a less highly elaborated form of a smaller nutritive value. This portion is described as non-albuminoid nitrogen, and in analyses of the present day the amount is always given as an additional source of information, although our knowledge of its exact value to the animal is rather uncertain.

The ultimate composition of the ash is also frequently determined, and examples of the results obtained are of interest as showing the mineral matter that grasses withdraw from the soil.

Without entering into a discussion of the nutritive value of the several constituents of the grasses, for which reference can be made to Armsby's Manual of Cattle Feeding, it is sufficient to say that during the past few years the greater portion of the species described by Dr. Vasey in the preceding portion of this Bulletin have been analyzed, and the results collected and re-arranged, with some corrections, from the annual reports of the Department are presented in the following pages.

The first series consists of analyses made with specimens collected at or near the time of blooming. Their origin is as follows:

No. of anal.

1. *Paspalum læve* (Water Grass). From Prof. S. B. Buckley, Austin, Tex. 1878.
2. *Paspalum læve* (Water Grass). From the Eastern Experimental Farm, West Grove, Chester County, Pa. 1880, August 23-29.
3. *Paspalum dilatatum.* From S. L. Goodale, Saco, Me. 1880.
4. *Paspalum praecox.* From Charles Mohr, Mobile, Ala. 1879.

No. of anal.

5. *Panicum filiforme*. From Charles Mohr, Mobile, Ala. 1878.
6. *Panicum sanguinale* (Crab Grass). From Charles Mohr, Mobile, Ala. 1878.
7. *Panicum sanguinale* (Crab Grass). From the grounds of the Department. June 23, 1880.
8. *Panicum sanguinale* (Crab Grass). From the Eastern Experimental Farm, West Grove, Chester County, Pa. August 11, 1880.
9. *Panicum marinum*. From Charles Mohr, Mobile, Ala. 1878.
10. *Panicum Texanum* (Texas Millet). From Prof. S. B. Buckley, Austin, Tex. 1879.
11. *Panicum proliferum* (Large Crab Grass). "Very ripe and rank." From Charles Mohr, Mobile, Ala. 1879.
12. *Panicum agrostoides* (Marsh Panic). From W. S. Robertson, Muscogee, Ind. T. 1879.
13. *Panicum anceps*. From Charles Mohr, Mobile, Ala. 1879.
14. *Panicum anceps*. From the Eastern Experiment Farm, West Grove, Chester County, Pa. July 31, 1880.
15. *Panicum Crus-galli* (Barnyard Grass). From Charles Mohr, Mobile, Ala. 1879.
16. *Panicum Crus-galli* (Barnyard Grass). From Prof. S. B. Buckley, Austin, Tex.
17. *Panicum Crus-galli* (Cock'sfoot). From the Eastern Experimental Farm, West Grove, Chester County, Pa. August 25, 1880.
18. *Panicum virgatum* (Panic Grass). From W. S. Robertson, Muscogee, Ind. T. 1879. Low growth.
19. *Panicum virgatum* (Tall Panic or Switch Grass). From Prof. S. B. Buckley, Austin, Tex. 1878.
20. *Panicum virgatum* (Tall Panic or Switch Grass). From Charles Mohr, Mobile, Ala. 1878.
21. *Panicum virgatum* (Tall Panic or Switch Grass). From W. S. Robertson, Muscogee, Ind. T. 1879. Tall growth.
22. *Panicum divaricatum*. From Charles Mohr, Mobile, Ala. 1879.
23. *Panicum gibbum*. From Charles Mohr, Mobile, Ala. 1879.
24. *Panicum obtusum*. From Prof. S. B. Buckley, Austin, Tex. 1878.
25. *Panicum capillare* (Witch Grass). From W, S. Robertson, Muscogee, Ind. T. 1879.
26. *Panicum dichotomum*. From Charles Mohr, Mobile, Ala. 1879.
27. *Setaria Italica*. From the Eastern Experimental Farm, West Grove, Chester County, Pa. July 24, 1880.
28. *Setaria glauca* (Fox-tail). From the grounds of the Department. July 24, 1880.
29. *Setaria glauca* (Fox-tail). From the Eastern Experimental Farm, West Grove, Chester County, Pa. August 11, 1880.
30. *Setaria setosa* (Bristle Grass). From Prof. S. B. Buckley, Austin, Tex. 1878.
31. *Tripsacum dactyloides* (Gama Grass). From D. L. Phares, Woodville, Miss. 1878.
32. *Tripsacum dactyloides* (Gama Grass). From the Eastern Experimental Farm, West Grove, Chester County, Pa. 1880.
33. *Spartina cynosuroides* (Whip Grass). From A. C. Lathrop, Glenwood, Pope County, Minnesota. 1879.
34. *Spartina cynosuroides* (Whip Grass). From J. D. Waldo, Quincy, Ill. 1879.
35. *Spartina cynosuroides* (Whip Grass). From W. S. Robertson, Muscogee, Ind. T. 1879.
36. *Andropogon Virginicus* (Brown Sedge, Sedge Grass). From Prof. S. B. Buckley, Austin, Tex. 1878.
37. *Andropogon scoparius*. From Charles Mohr, Mobile, Ala. 1879. Before bloom.
38. *Andropogon scoparius* (Broom Grass). From Charles Mohr, Mobile, Ala. 1878.
39. *Andropogon scoparius*. From W. S. Robertson, Muscogee, Ind. T. 1879.
40. *Andropogon macrourus* (Broom Grass). From Charles Mohr, Mobile, Ala.
41. *Andropogon furcatus* (Blue joint Grass). From A. C. Lathrop, Glenwood, Pope County, Minn. 1879.

No. of anal
42. *Andropogon furcatus* (Blue joint). From D. H. Wheeler, Nebraska. 1879.
43. *Andropogon furcatus* (Blue joint). From W. S. Robertson, Muscogee, Ind. T. 1879.
44. *Andropogon furcatus* (Blue joint Grass). From the Eastern Experimental Farm, West Grove, Chester County, Pa. September 2, 1880.
45. *Andropogon argenteus* (Silver Beard Grass). From W. S. Robertson, Muscogee, Ind. T. 1880.
46. *Sorghum halepense* (Johnson Grass). From Charles Mohr, Mobile, Ala. 1878.
47. *Sorghum nutans.* From W. S. Robertson, Muscogee, Ind. T. 1879.
48. *Sorghum nutans* (Wood Grass). From Prof. S. B. Buckley, Austin Tex. 1878.
49. *Phalaris intermedia,* var. *angusta* (American Canary Grass). From South Carolina, 1879.
50. *Anthoxanthum odoratum* (Sweet Vernal Grass). From James O. Adams, Manchester, N. H. 1879.
51. *Anthoxanthum odoratum* (Sweet Vernal Grass). From the Eastern Experimental Farm, West Grove, Chester County, Pa. May 11-24, 1880.
52. *Anthoxanthum odoratum* (Sweet Vernal Grass). From the grounds of the Department. May 1, 1880.
53. *Hierochloa borealis* (Vanilla Grass). From E. Hall, Athens, Ill. 1878.
54. *Alopecurus pratensis* (Meadow Fox-tail). From the grounds of the Department. May 1, 1880.
55. *Aristida purpurascens* (Purple Beard Grass). From W. S. Robertson, Muscogee, Ind. T.
56. *Milium effusum.* From C. G. Pringle, Hazen's Notch, Vt. 1880.
57. *Muhlenbergia diffusa* (Dropseed Grass). From Prof. S. B. Buckley, Austin, Tex. 1878.
58. *Muhlenbergia diffusa* (Dropseed Grass). From the Eastern Experimental Farm, West Grove, Chester Grove, Pa. Aug. 25, 1880.
59. *Muhlenbergia Mexicana.* From Eastern Experiment Farm, West Grove, Chester County, Pa. August 22, 1880.
60. *Muhlenbergia glomerata* (Satin Grass). From A. C. Lathrop, Glenwood, Pope County, Minn. 1879.
61. *Muhlenbergia sp.?* (Knot Grass). From James O. Adams, Manchester, N. H. 1879.
62. *Phleum pratense* (Timothy, Herd's Grass). From the grounds of the Department. June, 18, 1880.
63. *Phleum pratense* (Timothy). From the grounds of the Department. Wayside growth. June 4, 1880.
64. *Phleum pratense* (Timothy). From the grounds of the Department. June 26, 1882. First year's growth from seed.
65. *Phleum pratense* (Timothy). From the Eastern Experimental Farm, West Grove, Chester County, Pa. June 20, 1880.
66. *Phleum pratense* (Timothy, Herd's Grass). From J. W. Sanborn, Hanover, N. H. 1881.
67. *Phleum pratense* (Timothy). From W. H. Hackstaff, La Fayette, Ind. 1882.
68. *Phleum pratense* (Timothy). From J. M. Robinson, Queen Anne County, Md. July 4, 1882. Unmanured for years.
69. *Sporobolus Indicus* (Sweet Grass). From D. L. Phares, Woodville, Miss. 1878.
70. *Agrostis vulgaris* (Redtop, Herd's Grass). From the grounds of the Department June 23, 1880.
71. *Agrostis vulgaris* (Redtop). From the grounds of the Department. Wayside growth. June 18, 1880.
72. *Agrostis vulgaris* (Herd's Grass). From the Eastern Experimental Farm, West Grove, Chester County, Pa.
73. *Agrostis vulgaris* (Redtop). From J. J. Rosa, Milford, Del.
74. *Agrostis exarata* (Native Redtop). From Theo. Louis, Louisville, Wis. 1878.

No. of anal.

75. *Cinna arundinacea* (Reed Grass). From W S. Robertson, Muscogee, Ind. T.

76. *Holcus lanatus* (Velvet Grass). From the grounds of the Department. May 25, 1880.

77. *Arena striata* (Mountain Oat Grass). From Cyrus G. Pringle, Charlotte, Vt. 1879.

78. *Arrhenatherum arenaceum* (Oat Grass). From the grounds of the Department. May 25, 1880.

79. *Arrhenatherum arenaceum* (Oat Grass). From Dr. W. C. Benbow, Greensborough, N. C. Late bloom. May 12, 1880.

80. *Danthonia spicata* (Wild Oat Grass). From James O. Adams, Manchester, N. H. 1879.

81. *Danthonia compressa* (Wild Oat Grass). From Cyrus G. Pringle, Charlotte, Vt. 1847.

82. *Cynodon Dactylon* (Bermuda Grass) From Charles Mohr, Mobile, Ala. 1878.

83. *Cynodon Dactylon* (Bermuda Grass.) From D. L. Phares, Woodville, Miss. 1878.

84. *Bouteloua oligostachya* (Gramma Grass). From A. C. Lathrop, Glenwood, Pope County, Minn., 1879.

85. *Eleusine Indica* (Yard Grass, Crowfoot Grass). From Prof. S. B. Buckley, Austin, Tex.

86. *Eleusine Indica* (Yard Grass, etc.). From Dr. W. A. Cresswell, Americus, Ga. 1878.

87. *Eleusine Indica* (Yard Grass, etc.). From Charles Mohr, Mobile, Ala. 1878.

88. *Leptochloa mucronata* (Feather Grass). From Prof. S. B. Buckley, Austin, Tex. 1878.

89. *Triodia purpurea* (Sand Grass). From W. S. Robertson, Muscogee, Ind. T. 1879.

90. *Triodia seslerioides* (Tall Redtop). From Prof. S. B. Buckley, Austin, Tex. 1878.

91. *Uniola latifolia* (Fescue Grass). From Charles Mohr, Mobile, Ala. 1879.

92. *Uniola latifolia* (Fescue Grass). From M. S. Robertson, Muscogee, Ind. T. 1879.

93. *Dactylis glomerata* (Orchard Grass). From James O. Adams, Manchester, N. H. 1879.

94. *Dactylis glomerata* (Orchard Grass). From the grounds of the Department, May 13, 1880. First growth.

95. *Dactylis glomerata* (Orchard Grass). From the grounds of the Department, June 18, 1880. Later growth.

96. *Dactylis glomerata* (Orchard Grass). From the Eastern Experimental Farm, West Grove, Chester County, Pa. 1880.

97. *Dactylis glomerata* (Orchard Grass). From W. H. Cheek, Warren County, N. C. Early bloom. May 16, 1880.

98. *Dactylis glomerata* (Orchard Grass). From Dr. W. C. Benbow, Greensborough, N. C. 1880. Early bloom, May 12.

99. *Dactylis glomerata* (Orchard Grass). From S. L. Goodale, Saco, Me. 1880.

100. *Poa pratensis* (Blue Grass, June Grass). From Theo. Louis, Louisville, Wis. 1878.

101. *Poa pratensis* (Blue Grass, etc.) From James O. Adams, Manchester, N. H. 1879.

102. *Poa pratensis* (Blue Grass, etc.). From the grounds of the Department. May 28, 1880. Growth from best soil.

103. *Poa pratensis* (Blue Grass, etc.). From the grounds of the Department. May 8, 1880. Growth from poorer soil.

104. *Poa pratensis* (Blue Grass, etc.). From the grounds of the Department. May 19, 1880. Growth by wayside.

105. *Poa pratensis* (Blue Grass, etc.) From J. D. Waldo, Quincy, Ill. May 17, 1880.

106. *Poa pratensis* (Blue Grass, etc.). From W. B. Cheek, Warren County, N. C. 1880. Before bloom.

No. of anal.

107. *Poa pratensis* (Blue Grass, etc.). From the Eastern Experimental Farm, West Grove, Chester County, Pa. 1880.

108. *Poa compressa* (English Blue Grass, Wire Grass). From James O. Adams, Manchester, N. H. 1879.

109. *Poa compressa* (English Blue Grass). From the grounds of the Department. June 17, 1880.

110. *Poa compressa* (English Blue Grass). From the Eastern Experimental Farm, West Grove, Chester County, Pa. June 10, 1880.

111. *Poa compressa* (English Blue Grass). From J. J. Rosa, Milford, Del. June 6, 1880.

112. *Poa serotina* (Fowl Meadow Grass, False Redtop). From Theo. Louis, Louisville, Wis. 1878.

113. *Poa arachnifera.* From Ellis County, Tex. 1882.

114. *Poa alsodes.* From the Eastern Experimental Farm, West Grove, Chester, County, Pa. June 2, 1880.

115. *Glyceria aquatica* (Reed Meadow Grass). From Cyrus G. Pringle, Charlotte, Vt. 1879.

116. *Glyceria nervata* (Fowl Meadow Grass). From Cyrus G. Pringle, Charlotte, Vt. 1879.

117. *Glyceria nervata* (Fowl Meadow Grass). From James O. Adams, Manchester, N. H. 1879.

118. *Glyceria nervata* (Fowl Meadow Grass). From the Eastern Experimental Farm, West Grove, Chester County, Pa. June 2, 1880.

119. *Festuca ovina* (Sheep's Fescue). From James O. Adams, Manchester, N. H 1879.

120. *Festuca ovina* (Sheep's Fescue). From the grounds of the Department. May 21 1880.

121. *Festuca elatior* (Meadow Fescue). From the Eastern Experimental Farm, West Grove, Chester County, Pa. June 2, 1880.

122. *Festuca pratensis* (Meadow Fescue). From James O. Adams, Manchester, N. H 1879.

123. *Festuca pratensis* (Field Fescue). From the grounds of the Department. June 1, 1880. After bloom.

124. *Bromus secalinus* (Cheat, Chess). From James O. Adams, Manchester, N. H. 1879.

125. *Bromus unioloides* (Schrader's Grass). From the grounds of the Department. 1879.

126. *Bromus unioloides* (Schrader's Grass). From the grounds of the Department. May 13, 1880.

127. *Bromus erectus* (Chess). From the grounds of the Department. May 19, 1880.

128. *Bromus carinatus* (California Brown-grass). From E. Hall, Athens, Ill. 1878.

129. *Lolium perenne* (Common Darnel, Ray, or Rye Grass). From the grounds of the Department. June 1, 1880.

130. *Lolium perenne* (Rye Grass, etc.). From the Eastern Experimental Farm, West Grove, Chester County, Pa.

131. *Lolium Italicum* (Italian Rye Grass). From the grounds of the Department. May 26, 1882.

132. *Agropyrum repens* (Couch, Quitch, or Quack Grass). From James O. Adams. Manchester, N. H. 1879.

133. *Agropyrum repens* (Couch Grass, etc.). From the Eastern Experimental Farm, West Grove, Chester County, Pa.

134. *Agropyrum repens* (Couch Grass). From S. L. Goodale, Saco, Me. 1880.

135. *Agropyrum repens* (Couch Grass, etc.). From the grounds of the Department. June 23, 1880. Early bloom.

136. *Elymus canadensis* (Wild Rye Grass). From D. H. Wheeler, Nebraska. 1879.

The specimens, it will be seen, are from all parts of the country and grown under every condition of soil and environment. Those collected by Dr. Peter Collier in 1878 and 1879 were mostly from the poorer soils, and were intended to represent the wild grasses of the country. Those collected in subsequent years by myself were chiefly cultivated varieties. The development in nearly every case was full bloom or shortly after, that being the period at which the grasses as a whole seem to be cut for hay.

The analyses have been calculated for " dry substance," and also for "fresh grass," where the amount of water in the fresh grass had been determined; otherwise, for the average amount of water in hay as given by Wolff. This figure is probably too high for the United States, owing to our drier climate; but, in the absence of exact data for the selection of a more accurate one, it has been provisionally accepted. It is very easy to calculate from the composition of the dry substance what effect the presence of any percentage of water would have on the absolute amount of any constituent present in a given weight of grass.

No. of analysis	Plant	Specimen No.	Date of cutting	Locality	Height in inches	Ash.	Fat.	Nitrogen-free extract.	Crude fiber.	Albuminoids.	Total nitrogen.	Non-albuminoid.	Per cent. of nitrogen as non-albuminoid.	Water.	Ash.	Fat.	Nitrogen-free extract.	Crude fiber.	Albuminoids.
1	Paspalum laeve	1	1878.	Tex		7.20	2.75	53.83	27.61	8.11	1.30	.62	47.5	14.30	6.60	2.36	46.13	23.66	6.95
2	do.	2	1880.	Pa		6.90	2.16	59.27	23.50	8.17	1.31	.38	29.0	14.30	5.91	1.85	50.50	20.14	7.00
3	Paspalum dilatatum	3	Aug. 23-29	Me		8.49	2.21	58.42	25.31	6.13	.98	.28	26.5	14.30	7.28	1.89	50.07	21.21	5.25
4	Paspalum precox	4		Ala		7.41	3.60	57.75	25.31	6.93	.95	.25	26.2	14.30	6.35	3.09	49.49	21.69	5.08
5	Panicum filiforme	5	1879.	Pa		8.76	1.48	60.95	32.09	2.98	.48	.00	0.6	14.30	7.57	1.27	52.24	22.14	2.54
6	Panicum sanguinale	6		Ala		12.61	2.82	42.70	32.90	9.78	1.57	.51	32.6	14.30	10.81	2.42	36.59	27.50	8.38
7	do.	7	June 23,'80	D.C		15.01	4.84	37.99	19.03	23.13	3.70			76.50	3.53	2.13	8.03	4.47	5.44
8	do.	8	Aug. 11,'80	Pa		11.45	3.26	50.56	22.90	11.83	1.89	.81	42.6	14.30	9.81	1.34	43.33	19.63	10.14
9	Panicum maximum	9	1878.	Tex		10.10	1.57	48.93	31.52	8.89	1.42	.74	62.0	14.30	7.75	2.79	41.98	27.01	7.62
10	Panicum Texanum	10	1878.	Ala		10.17	2.47	54.93	27.02	5.48	.88	.33	37.5	14.30	8.65	2.12	47.07	33.16	4.70
11	Panicum proliferum	11	1879.	Ind. T		6.69	3.01	50.87	30.86	5.89	.94	.63	35.7	14.30	9.58	2.88	43.42	20.03	9.49
12	Panicum agrostoides	12		Ala		9.05	5.69	50.87	30.87	5.78	1.24	.39	41.1	14.30	5.73	4.88	43.59	28.45	5.05
13	Panicum anceps	13		Pa		5.68	1.82	62.59	20.88	9.03	1.24	.19	30.7	14.30	7.47	1.50	43.36	21.95	4.95
14	do.	14	July 31	Ala		16.07	2.04	46.77	31.13	7.77	1.65	.45	39.7	14.30	1.50	1.84	47.46	17.80	5.74
15	Panicum Crus-galli	15		Pa		11.82	2.49	47.10	33.22	3.99	2.02	.15	38.1	14.30	5.98	1.73	53.64	24.78	6.66
16	do.	16	Aug. 25	Tex		7.24	1.66	49.39	36.78	3.80	2.00	.80	39.3	14.30	13.77	2.13	46.44	26.66	3.42
17	do.	17		Ind. T		4.57	1.92	41.80	28.82	4.93	.78	.49	22.3	14.30	10.13	1.42	40.08	3.69	10.80
18	Panicum virgatum	18		Tex			2.98	54.95	32.25	4.57	.73	.18	41.8	14.30	4.70	1.65	40.95	24.68	4.39
19	do.	19		Ind. T		14.29	4.16	28.82	24.16	2.80	.45	.21	47.6	14.30	6.20	2.55	48.81	21.09	4.23
20	do.	20		Ala		8.53	3.89	57.40	33.12	9.22	1.48	.36	24.6	14.30	3.92	2.16	42.33	24.05	3.92
21	Panicum divaricatum	21	1878.	Tex		13.12	3.55	50.93	28.24	12.22	1.96	.83	42.6	14.30	3.20	1.93	52.23	31.52	2.40
22	Panicum gibbum	22		Ala		6.50	2.71	46.44	29.44	7.24	1.19	.05	45.9	14.30	12.25	3.56	49.19	27.64	7.92
23	Panicum obtusum	23		Ind. T		10.13	2.00	55.30	24.52	6.94	1.12	.64	37.5	14.30	7.31	3.34	40.18	23.19	10.47
24	Panicum capillare	24	24 July	Ala		7.50	1.49	50.07	25.75	6.77	1.08	.50	33.6	14.30	9.38	3.04	43.65	20.71	6.21
25	Panicum dichotomum	25		D.C		7.27	2.39	55.78	21.94	9.49	1.53	.51	28.5	14.30	4.89	2.32	39.80	28.38	5.98
26	Setaria Italica	26		Pa		7.94		55.38	32.30	8.52	1.44	.41	58.7	14.30	8.68	.84	47.39	21.20	5.80
27	Setaria glauca	27	24 July	Tex		9.04		58.54	26.51	8.40	1.36	.39	41.1	14.30	6.43	2.62	42.91	21.02	8.13
28	do.	28	11 Aug.	Miss		6.19		56.31		8.60	1.37	.56	27.6	68.40	2.30	1.28	47.80	8.13	2.86
29	Setaria setosa	29										.38		14.30	.60	2.05	50.18	18.80	7.30
30	do.	30												14.30	7.78		41.86	27.68	7.28
31	Tripsacum dactyloides	31												14.30	6.30		48.26	22.72	7.37

No. of analysis	Specimen No.	Date of cutting	Locality	Height in inches	Dry substance								Fresh substance or hay					
					Ash	Fat	Nitrogen-free extract	Crude fiber	Albuminoids	Total nitrogen	Non-albuminoid	Per cent. of nitrogen as non-albuminoid	Water	Ash	Fat	Nitrogen-free extract	Crude fiber	Albuminoids
32	32	Aug. 11	Pa.		5.34	3.47	60.89	22.45	7.94	1.27	.32	25.2	14.30	4.58	2.97	52.11	19.24	6.80
33	33		Minn		7.22	3.42	51.76	25.79	9.81	1.57	.43	33.8	14.30	6.19	2.93	46.07	22.10	8.41
34	34		Ill		6.55	2.96	61.12	22.89	6.48	1.04	.27	25.5	14.20	5.61	2.54	52.38	19.62	4.55
35	35		Ind. T.		5.20	2.40	59.32	27.20	4.88	.78	.45	57.5	14.30	4.46	2.91	50.84	23.31	4.18
36	36		Tex.		9.33	1.67	52.92	33.08	3.00	.48	.12	24.4	14.30	8.00	1.59	43.35	28.35	2.57
37	37		Ala.		7.11	1.85	53.62	27.07	6.45	1.04	.26	24.7	14.30	6.09	1.59	47.58	24.91	5.53
38	38		Ind. T.		4.46	3.19	54.37	21.64	6.32	.90	.39	56.6	14.30	3.82	2.73	53.39	21.12	5.84
39	39		Ala.		5.09	3.02	58.19	29.44	4.14	.66	.53	59.7	14.30	3.21	2.18	50.03	25.57	3.94
40	40		Ind. T.		5.77	3.19	58.46	23.65	5.77	.92	.44	57.1	14.30	4.36	2.59	49.87	25.50	4.94
41	41		D. C.		5.09	2.54	57.60	25.26	8.05	1.29	.33	33.9	14.30	6.74	1.80	50.10	21.98	6.90
42	42		Nebr		4.08	3.19	61.10	25.38	4.32	.63	.38	37.5	14.30	3.50	2.73	49.36	22.50	4.56
43	43		Ind. T.		13.53	2.08	52.24	27.04	3.95	.63	.33	52.5	14.30	11.60	2.12	44.53	26.72	2.39
44	44		Ind. T.		8.07	2.64	61.19	23.39	4.99	.80	.09	11.2	14.30	6.92	2.62	55.01	23.17	4.28
45	45	Sept. 2	Ind. T.		9.17	2.63	52.25	25.05	3.73	.60	.36	60.7	14.30	4.46	2.43	44.77	21.76	3.20
46	46		Ind. T.		11.66	4.11	58.75	26.62	11.80	1.89	.76	40.0	14.30	7.86	2.18	51.21	21.47	10.11
47	47		Tex.		8.43	2.97	60.32	24.84	3.20	.51	.38	52.5	14.30	9.90	1.40	43.12	24.63	3.32
48	48		S. C.	15	5.83	2.54	35.70	35.79	15.95	2.55	.21	41.2	14.30	9.22	3.52	37.33	30.58	2.74
49	49		N. H.		9.32	2.97	53.81	25.49	8.56	1.37	.82	38.1	78.80	5.00	2.92	46.12	21.29	13.67
50	50	May 11–24	Pa.	60	7.75	4.08	59.45	20.02	9.47	1.52	.66	35.9	60.00	1.50	2.54	46.46	22.10	7.34
51	51	May 1	Ill		9.28	4.06	61.81	22.78	14.15	2.27	.15	37.7	14.30	1.99	.71	12.81	21.85	9.85
52	52		D. C.		9.28	2.59	54.36	22.78	10.81	1.73	.86	29.8	14.30	3.10	3.48	56.11	19.73	2.01
53	53		Ind. T.		16.49	3.87	46.33	24.55	15.97	1.69	.00	28.8	14.30	7.65	3.22	42.38	9.51	12.12
54	54	Aug. 25	Vt.		4.33	3.81	55.35	22.56	10.00	2.64	.63	17.0	14.30	5.65	3.23	21.72	21.32	4.53
55	55	Aug. 23	Pa.		15.02	3.43	47.35	21.94	10.88	1.74	.29	23.4	14.30	14.06	2.04	38.69	21.05	3.70
56	56		Pa.		6.34	2.69	65.47	22.69	4.32	.70	.18	28.5	14.30	56.11	2.30	47.44	20.19	13.69
57	57		Minn		5.66	5.77	41.21	17.68	20.32	3.25	.80	37.2	14.30	3.71	4.94	40.58	20.60	13.57
58	58		N. H.		6.56	4.73	53.76	21.77	13.40	2.15	.38	28.1	14.30	12.87	4.20	35.32	18.45	8.22
59	59	June 18	D. C.	58		3.58	58.93	21.93	9.90	1.58	.30	24.0	67.20	5.43	2.30	46.07	18.80	4.13
60	60	June 4	D. C.	60		3.95	57.48	24.53	8.48	1.30	.30	22.0	63.40	2.40	1.45	21.04	8.61	3.10

Plant names:

32 Tripsacum dactyloides
33 Spartina cynosuroides
34 ...do...
35 Andropogon Virginicus
36 Andropogon scoparius
37 ...do...
38 ...do...
39 ...do...
40 Andropogon macrourus
41 Andropogon furcatus
42 ...do...
43 ...do...
44 ...do...
45 Andropogon argenteus
46 Sorghum halepense
47 Sorghum nutans
48 ...do...
49 Phalaris intermedia var. angusta
50 Anthoxanthum odoratum
51 ...do...
52 ...do...
53 Hierochloa borealis
54 Alopecurus pratensis
55 Aristida purpurascens
56 Milium effusum
57 Muhlenbergia diffusa
58 ...do...
59 Muhlenbergia Mexicana
60 Muhlenbergia glomerata
61 Muhlenbergia sp. (?)
62 Phleum pratense. (?)
63 ...do...

No.	Species	Date	Loc.															
64	Phleum pratense	June 20	D. C		7.16	50.03	4.47	27.35	10.99	1.75	.51	29.1	66.75	2.38	1.49	16.64	9.09	3.65
65	do		Pa		5.05	57.52	3.22	23.39	9.12	1.44	.10	29.5	14.30	4.33	2.77	43.06	21.72	7.82
66	do		N. H		4.57	57.16	3.20	28.28	5.53	.93	.10	10.8	14.30	3.92	3.60	48.99	24.23	4.96
67	do		Ind		7.05	52.99	2.18	30.43	5.69	.88	.05	0.0	64.00	6.04	1.87	45.41	27.65	4.73
68	do	July 4, '82	Md	45	4.93	52.73	2.93	22.67	12.35	1.98	.57	12.2	14.30	1.74	1.48	19.02	10.98	2.74
69	Sporobolus Indicus		Miss	58.8	7.04	56.82	3.27	22.02	11.02	1.93	.53	28.7	61.40	8.03	2.80	41.28	22.00	10.35
70	Agrostis vulgaris	June 23	D. C		7.27	56.30	3.71	20.42	0.95	1.70	.80	30.1	58.80	2.81	1.11	21.83	8.50	4.29
71	do	June 18	Pa		5.84	58.49	3.31	25.33	9.00	1.50	.80	20.1	61.40	2.40	1.11	21.83	8.41	4.11
72	do	July 1	Del		8.88	34.58	2.30	21.68	11.25	1.58	.45	50.7	58.80	5.90	2.18	46.77	21.71	8.48
73	do	1882.	Wis		0.53	56.80	2.06	24.51	10.61	1.80	.47	25.0	14.30	5.10	2.84	46.66	18.75	9.64
74	Agrostis exarata		Ind. T		6.09	55.63	2.84	29.64	6.22	1.00	.00	26.3	14.30	5.73	1.97	48.53	21.01	0.09
75	Cinna arundinacea	May 25	D. C		8.43	55.47	2.06	26.16	7.35	1.30	.60	47.4	50.60	4.07	1.92	46.69	25.40	3.63
76	Holcus lanatus		Vt		4.96	55.53	3.89	24.33	8.75	1.40	.45	40.2	14.30	4.25	2.55	47.43	12.35	3.63
77	Avena striata	May 25	D. C		7.93	49.97	4.03	28.00	8.78	1.49	.47	0.0	62.30	2.99	1.92	22.42	22.42	7.50
78	Arrhenatherum avenaceum	May 12	N. C		4.38	56.92	2.81	28.11	12.79	2.12	.15	51.4	14.30	3.75	1.43	20.71	24.82	3.31
79	do		N. H		3.57	54.61	3.80	28.00	7.98	1.29	.09	13.1	14.30	3.02	1.53	42.82	24.36	10.88
80	Danthonia spicata		Vt		9.91	54.61	3.52	30.32	13.42	2.15	.32	22.0	14.30	7.81	1.26	24.78	24.95	4.96
81	Danthonia compressa		Ala		7.81	53.75	1.57	23.23	10.09	1.77	.30	17.4	14.30	7.81	3.02	45.09	24.98	6.84
82	Cynodon Dactylon		Miss		19.24	53.75	2.13	23.65	8.58	1.71	.01	36.0	14.30	8.1	1.34	20.16	19.96	11.50
83	do		D. C	72	8.24	31.01	3.12	22.49	13.00	1.84	.75	59.2	14.30	6.09	1.83	49.58	20.16	9.16
84	Bouteloua oligostachya		Tex		9.71	34.01	2.14	22.49	8.41	1.34	.81	8.7	14.30	8.7	2.67	29.25	19.41	7.35
85	Eleusine Indica		Ala		5.16	45.47	2.08	30.17	9.53	1.53	.59	32.7	14.30	4.83	1.78	47.91	26.58	9.16
86	Uniola latifolia		Tex		5.16	53.94	2.53	27.74	12.51	1.99	.81	47.8	77.30	10.08	2.17	46.92	19.27	11.65
87	do		Ind. T	85	14.64	54.61	2.05	21.74	9.01	1.37	.49	46.3	14.30	4.42	3.18	12.20	47.54	10.39
88	Leptochloa mucronata		Tex	87	10.44	36.01	1.99	27.51	10.29	1.58	.36	48.9	14.30	4.40	1.73	16.62	27.20	9.48
89	Triodia purpurea		Ala		8.07	52.40	3.12	27.94	8.74	1.41	.50	21.6	14.30	12.55	2.67	47.08	32.33	0.66
90	Triodia seslerioides		Ind. T		8.64	57.04	2.86	34.96	7.50	1.20	.09	30.3	77.00	5.93	2.99	44.70	21.13	0.90
91	Uniola latifolia	May 13	N. H		6.33	51.43	3.24	23.40	12.61	2.01	.02	3.1	14.30	7.23	.74	44.96	21.35	5.40
92	do	June 18	D. C		7.42	51.45	2.66	20.64	7.82	1.99	.19	8.4	69.00	1.83	1.32	57.04	5.77	9.33
93	Dactylis glomerata		Pa		8.90	55.32	2.66	27.51	9.01	1.37	.51	37.2	14.30	2.86	1.15	14.43	23.38	5.55
94	do		N. C	87	8.02	54.14	2.08	21.97	9.01	1.58	.47	19.0	14.30	6.36	1.23	17.62	23.38	7.21
95	do		D. C	80	5.20	36.01	2.39	20.97	8.74	1.41	.36	30.1	14.30	7.63	3.73	18.70	19.78	2.16
96	do		Pa		5.47	52.16	2.86	26.05	11.51	1.85	.50	25.7	14.30	6.87	3.93	41.47	21.40	4.14
97	do		N. C		8.30	57.04	4.95	27.94	7.50	1.20	.47	28.8	14.30	4.46	1.92	41.09	22.31	7.34
98	do		Ill	70	7.03	34.18	2.83	25.46	12.61	2.01	.38	30.3	69.00	4.69	2.95	49.00	23.94	8.49
99	do		Me		7.73	34.13	3.77	21.10	15.09	1.67	.43	1.0	14.30	2.33	3.03	17.05	23.95	8.82
100	Poa pratensis	May 21	Wis		7.82	47.95	4.58	24.13	14.22	2.41	.41	8.4	70.70	2.18	3.64	47.06	6.64	7.49
101	do	June 8	N. H		9.01	31.18	4.32	20.64	1.51	2.03	.54	19.3	14.30	2.61	3.29	47.49	7.81	9.89
102	do	May 19	D. C	65	5.21	55.33	4.35	23.35	7.82	2.27	.47	20.8	14.30	6.70	1.92	50.11	21.30	6.43
103	do	May 17	D. C	70	6.08	53.41	3.77	21.91	8.83	1.41	.38	25.2	14.30	6.41	3.71	47.06	21.68	3.54
104	do	June 19	Del		7.85	55.03	3.84	21.95	12.69	2.33	.43	26.5	70.70	1.78	2.95	56.49	21.73	2.43
105	do	June 17	Wis		6.94	55.01	4.25	20.15	13.45	2.15	.38	6.5	14.30	1.73	3.18	30.13	5.41	3.53
106	do	June 16	Tex		4.23	47.16	2.84	20.35	6.27	1.01	.63	6.8	42.10	3.53	3.06	44.06	17.27	12.48
107	do		Pa		2.84	51.41	4.09	21.89	13.84	2.21	.33	33.	14.30	0.96	3.51	46.64	17.33	12.18
108	Poa compressa		D. C	30	11.62	51.16	2.30	20.35	8.13	1.30	.30	22.7	11.30	6.26	1.89	46.64	21.94	1.56
109	do		Del		9.41													3.72
110	do		Wis		7.30													7.66
111	do		Tex															11.53
112	Poa serotina		Pa															5.37
113	do		V.	2														8.10
114	Poa arachnifera	June																11.68
115	Glyceria aquatica																	6.97

No. of analysis	(species)	Specimen No.	Date of cutting	Locality	Height in inches	Dry substance Ash	Fat	Nitrogen-free extract	Crude fiber	Albuminoids	Total nitrogen	Non-albuminoid	Per cent. of nitrogen as non-Albuminoid	Water	Fresh Ash	Fat	Nitrogen-free extract	Crude fiber	Albuminoids
116	Glyceria nervata	116		Vt.		6.19	3.20	53.00	28.20	9.41	1.50	.33	22.1	14.30	5.30	2.74	45.43	21.17	8.06
117	do	117	June	N.H.		6.80	2.91	60.01	21.97	8.31	1.31	.45	34.1	14.30	5.83	2.49	51.43	18.62	7.12
118	do	118		Pa.		7.03	2.87	53.01	22.38	14.83	2.37	.58	24.6	14.30	6.79	2.46	45.43	18.32	12.70
119	Festuca ovina	119		N.H.	40	5.03	4.26	81.18		6.53	1.04	.16	10.5	67.00	4.31	3.05	72.14	7.85	5.60
120	do	120	May 21	D.C.		5.60	2.51	58.20	23.79	9.90	1.52	.77	35.0	14.30	1.85	.83	19.21	19.29	3.26
121	Festuca elatior	121	May 2	Pa.		6.07	3.28	51.59	25.50	13.77	2.20	.79	44.9	14.30	6.91	3.48	44.22	20.78	11.80
122	Festuca pratensis	122	June	N.H.	76	7.16	3.80	52.59	24.25	10.75	1.72	.57	43.5	14.30	7.83	2.81	45.07	23.68	9.74
123	do	123	June 1	N.H.		7.13	3.58	50.34	27.63	11.37	1.82	.96	46.4	14.30	6.14	3.49	43.31	20.39	9.21
124	Bromus secalinus	124		N.H.		7.12	4.08	57.30	23.79	13.62	2.18	.34	4.2	14.30	6.10	3.07	44.97	17.64	23.68
125	Bromus unioloides	125		D.C.	78	9.26	3.90	52.47	20.50	12.63	2.02	.34	16.8	14.30	8.35	.82	49.11	4.67	6.61
126	do	126	May 13	D.C.	68	7.70	2.81	56.19	24.52	12.69	2.41	.34	24.1	70.40	1.91	1.02	10.60	22.91	11.67
127	Bromus erectus	127	May 19	Ill.		10.88	2.68	49.79	20.73	9.02	1.59	.47	21.2	63.70	2.79	2.30	42.67	8.30	2.60
128	Bromus carinatus	128		D.C.	52	7.50	2.64	56.84	25.43	7.60	1.21	.41	33.1	14.30	9.32	.97	20.97	20.97	3.19
129	Lolium perenne	129	June 1	D.C.	90	6.09	2.32	63.94	18.26	8.87	1.42	.45	7.8	14.30	2.77	.51	51.89	15.65	2.81
130	do	130	May 26	Pa.		11.02	3.51	61.73	20.44	14.49	1.84	.18	24.5	78.00	5.22	2.43	11.39	11.39	8.50
131	Lolium Italicum	131	May	N.H.		9.32	3.62	56.27	19.41	11.48	1.35	.62	15.9	14.30	2.43	.51	48.22	4.50	3.16
132	do	132	June 12	Pa.		6.23	3.83	56.95	24.08	8.43	1.84	.60	29.7	14.30	7.99	3.02	49.50	16.63	7.00
133	Agropyrum repens	133		Me.		7.28	3.36	50.95	25.30	12.64	2.02	.26	18.7	14.30	6.34	3.00	43.66	20.64	9.84
134	do	134		D.C.		8.77	3.36	59.37	19.70	8.90	1.41	.30	38.3	58.30	6.25	3.28	24.76	21.68	10.83
135	do	135	June 23	Ind. T.		5.99	3.71	50.78	31.66	4.80	.77			14.30	3.65	1.40	52.59	8.22	3.67
136	Elymus Canadensis	136												14.30	5.87	2.22		21.32	3.70

The great variation in composition of grasses becomes apparent on examining the one hundred and thirty-six analyses; and by selecting the highest and lowest determinations the following table of extremes has been prepared:

Limits of composition of grasses.

Dry substance.	Highest.	Lowest.
Ash	19.24	3.57
Fat	5.77	1.48
Nitrogen-free extract	66.01	34.01
Crude fiber	37.72	17.68
Albuminoids	24.13	2.80
Nitrogen	3.70	.45
Non-albuminoid nitrogen	1.64	.00
Per cent. of nitrogen as non-albuminoid	60.70	.00
Water in fresh grass	76.50	60.00

The highest ash is undoubtedly owing to the presence of adherent soil, and the lowest carbo-hydrates are dependent relatively on the same cause. The wide variations in fiber and albuminoids must be regarded, however, as being entirely due to physiological causes, which are difficult to explain. *Panicum sanguinale*, for instance, which in one specimen contains the extreme amount of albuminoids and a small amount of fiber has in another only half as much albumen and one and three-quarter times as much fiber. We learn, then, that species are not in themselves at all fixed in their composition, there being as large variations among specimens of the same as between specimens of different species. Examples may be found in several portions of the preceding tables, and, for illustration, several analyses of *Phleum pratense* and of *Dactylis glomerata* from widely separated localities are given in the following tables:

Analyses of Phleum pratense (Timothy) from various localities.

FULL BLOOM.

Locality.	Ash.	Fat.	Nitrogen-free extract.	Crude fiber.	Albuminoids.	Total nitrogen.	Non-albuminoid nitrogen.	Per cent. of total nitrogen as non-albuminoids.
Department garden, 1881	7.16	4.47	50.03	27.35	10.90	1.75	.51	29.1
Department garden, 1880	5.64	3.58	58.93	21.93	9.90	1.58	.38	24.0
Maryland	4.93	4.22	52.83	30.43	7.60	1.23	.15	12.2
New Hampshire	4.57	4.20	57.16	28.28	5.79	.91	.10	10.8
Indiana	7.05	2.18	52.99	32.26	5.52	.88	.00	.0

Analyses of Dactylis glomerata (Orchard grass) from various localities

FULL BLOOM.

Locality.	Ash.	Fat.	Nitrogen-free extract.	Crude fiber.	Albuminoids.	Total nitrogen.	Non-albuminoids, nitrogen.	Per cent of total nitrogen as non-albuminoids.
North Carolina	7.42	3.56	56.03	23.08	9.91	1.58	.30	19.0
District of Columbia	8.07	3.24	53.76	25.40	9.53	1.53	.16	10.5
Maine	8.02	3.39	54.80	26.05	8.74	1.40	.36	25.7
District of Columbia	6.00	3.62	57.34	24.42	8.62	1.38	.42	30.4
Pennsylvania	6.33	2.66	54.90	27.51	8.56	1.37	.51	37.2
New Hampshire	8.44	3.49	54.75	24.91	8.41	1.35	.42	30.9

AVERAGE.

	Ash.	Fat.	Nitrogen-free extract.	Crude fiber.	Albuminoids.	Total nitrogen.	Non-albuminoids, nitrogen.	Per cent of total nitrogen as non-albuminoids.
Five localities	7.38	3.33	55.17	25.19	8.91	1.43	.36	25.2

The average composition of American grasses, as derived from the preceding analyses of grasses in bloom, and averages for different sections of the country, has been calculated. The results presented in the following table, with Wolff's averages for German grasses, are of interest:

Average composition of grasses.

	Number of analyses.	Ash.	Fat.	Nitrogen-free extract.	Crude fiber.	Albuminoids.	Total nitrogen.	Non-albuminoids, nitrogen.	Per cent of total nitrogen as non-albuminoids.
American :									
United States	135	7.97	3.14	53.97	25.71	9.21	1.47	.45	30.6
North of Potomac	70	7.64	3.44	53.01	23.70	10.21	1.63	.32	19.6
South	27	8.80	2.74	52.55	26.68	9.23	1.47	.56	38.1
Middle West	8	7.12	2.96	54.58	25.39	9.95	1.60	.41	25.0
West of Mississippi	30	8.23	2.86	52.67	29.60	6.64	1.06	.41	38.7
German (Wolff) :									
Fair		6.30	2.34	46.53	34.09	10.74	1.72		
Good		7.23	2.92	47.84	30.69	11.32	1.81		
Very good		8.24	3.29	48.93	25.77	13.77	2.20		

The different sections furnish very different qualities of grasses, and for the reason that those from the North were almost entirely from cultivated soil, while those from the other sections were many or most of them wild species from old sod. The improvement brought about by cultivation is marked, and the difference between a ton of wild Western and Eastern cultivated hay is apparent.

In comparison with German grasses our best do not equal in amount of albuminoids those classed by Wolff as *fair*, but they are far superior in having a much smaller percentage of fiber and consequently a larger amount of digestible carbo-hydrates. In the grasses of both countries the fiber increases with regularity as the nitrogenous constituents decrease, and of the latter the non-albuminoid portion is relatively great the poorer the quality of the grass.

CHANGES IN COMPOSITION DURING GROWTH.

In addition to the single analysis previously tabulated, analyses have been made of series illustrating the changes in composition of several species from the appearance of the blade to the maturity of the seed.

The grasses examined comprise :

I. *Agrostis vulgaris.* (Redtop.)
   A series from richer soil.
   A series from poorer soil.
II. *Phleum pratense.* (Timothy.)
   A series from richer soil.
   A series from poorer soil.
   A series of first year's growth from seed sown in garden soil.
   A series from La Fayette, Ind.
   A series from Hanover, N. H., the two latter from rather poor soil.
III. *Dactylis glomerata.* (Orchard Grass.)
   A series from the first growth.
   A series from later growth.
IV. *Alopecurus pratensis.* (Meadow Fox tail.)
   A series from good sod.
V. *Poa pratenses.* (Blue Grass, Meadow Grass.)
   A series from good soil.
   A series from poorer soil.
   A series from the wayside.
   A series from Quincy, Ill.
VI. *Poa compressa.* (Wire Grass.)
   A series from poor soil.
VII. *Bromus unioloides.* (Schrader's Grass.)
   A series from rich, garden soil.
VIII. *Bromus erectus.* (Broom Grass.)
   A series from poor soil.
IX. *Holcus lanatus.* (Satin Grass.)
   A series from poor soil.
X. *Arrhenatherum avenaceum.*
   A series from medium soil.
XI. *Setaria glauca.*
   A series from medium soil.
XII. *Anthoxanthum odoratum.* (Sweet Vernal Grass.)
   A series from medium soil.
XIII. *Festuca ovina.* (Sheep's Fescue.)
   A series from poor soil, growing in bunches.
XIV. *Lolium perenne var. Italicum.* (Italian Rye Grass.)
   A series from low, moist soil.
   A series of first year's growth from the seed in garden soil.
XV. *Lolium perenne.* (Rye Grass, Darnel.)
   A series from medium soil.

With a few exceptions the specimens were personally collected in the grounds of the Department, and are to be so understood when nothing else is said in their description. They all grew in the summer of 1880 except the few series illustrative of the first year's growth of certain species. The character of the soils has been designated as rich or garden soil, good soil, poorer soil, and wayside soil. The first is that of the experimental garden of the Department, and is above the average richness of cultivated soils; the second is that of the lawns about the Department building, the third, a light gravelly soil, occurring in a portion of the grounds, and the last the gutters and paths.

The specimens were cut close to the roots, weighed and dried rapidly in a current of air at 60° C. The methods of analysis were such as have been described in previous reports.

Description	When cut	Height in centimeters	Albuminoids	Crude fiber	Nitrogen-free extract	Fat	Ash	Water	Per cent. of total nitrogen, non-albuminoid	Non-albuminoid old nitrogen	Total nitro.	Albuminoids	Crude fiber	Nitrogen-free extract	Fat	Ash
I.—AGROSTIS VULGARIS.																
DEPARTMENT GROUNDS.																
Good soil:																
Panicle not out	June 1	42	4.25	6.75	17.35	1.21	2.64	67.8	38.9	.82	2.11	13.19	20.97	53.88	3.77	8.19
Panicle out; closed	June 1	58	4.34	6.66	17.27	1.29	2.54	68.1	36.7	.90	2.18	13.61	20.87	54.13	4.05	7.34
In early bloom	June 10	48	3.81	6.47	16.29	1.11	2.25	70.1	26.4	.54	1.76	11.02	21.64	54.46	3.02	7.53
In full bloom	June 23	45	4.25	8.50	21.04	1.08	2.80	61.4	30.1	.53	2.04	12.73	22.02	56.82	2.87	7.27
Seed in the milk	July 1	43	4.88	9.07	28.03	1.64	3.08	53.3	21.6	.30	1.67	10.44	28.03	60.02	3.51	7.60
Seed hard	July 5	47	4.59	10.02	28.56	2.06	3.27	51.5	11.8	.18	1.52	9.47	20.66	58.88	4.25	6.04
Seed mature	July	55	3.82	9.35	26.37	1.18	2.28	57.0	6.3	.00	1.43	8.80	21.75	61.32	2.74	5.30
Poorer soil:																
Panicle spreading	June 16	43	3.12	6.52	18.26	1.23	2.67	68.2	17.8	.28	1.57	9.81	20.49	57.41	3.88	8.41
Early blooming	June 18	53	4.10	8.41	24.10	2.18	2.41	56.8	20.1	.32	1.59	9.05	20.44	58.49	5.30	5.84
II.—PHLEUM PRATENSE.																
DEPARTMENT GROUNDS.																
Good soil:																
Spike invisible	June 1	42	3.67	5.83	15.92	1.34	2.54	70.7	35.0	1.70	2.01	12.54	19.91	54.31	4.56	8.68
Spike visible	June 1	62	3.34	5.91	16.40	.96	1.80	71.9	29.5	.55	1.86	11.90	21.03	57.20	3.40	6.41
Before bloom	June 23	62	3.36	7.16	17.61	1.18	2.19	67.5	21.8	.18	1.65	10.33		54.19	3.63	9.62
In early bloom	June 23	60	3.58	7.97	20.06	1.33	2.12	64.9	18.4	.38	1.58	9.90		58.91	3.85	6.04
In full bloom	June 18	58	3.25	7.19	19.33	1.17	1.86	67.2	24.0	.51		12.10			3.56	6.66
Early seed	June 18	52	2.68	8.08	21.34	.70	2.34	77.8	26.4		1.93			51.07	3.40	10.33
Poorer soil:																
In bloom	June 1	60	3.10	8.61	21.04	1.45	2.40	63.4	22.0	.30	1.36	8.44	23.53	57.46	3.05	6.56
In full bloom	July 1	70	2.10	6.42	17.10	.84	1.58	71.9	30.3	.30	1.19	7.46	22.84	61.08	2.98	5.64
DEPARTMENT GARDEN.																
First year's growth:																
Head out	June 19	49	3.03	5.14	10.12	1.31	1.84	78.50	17.3	.39	2.26	14.15	31.95	47.22	6.10	8.58
In bloom	June 26	78	3.65	9.09	16.64	1.49	2.38	60.75	29.1	.51	1.75	10.99	27.35	50.03	4.47	7.16
After bloom	July 3	65	3.79	12.26	22.46	2.03	2.63	56.63	17.9	.25	1.40	8.74	28.26	51.79	4.69	6.52
Do	July 10	75	3.37	11.14	22.79	1.53	2.31	58.86	11.3	.15	1.27	8.1e	27.08	55.39	3.72	5.63

II.—PHLEUM PRATENSE—Continued.

Description	When cut	Height in centimeters	Ash	Fat	Nitrogen-free extract	Crude fiber	Albuminoids	Total nitrogen	Non-albuminoid old nitrogen	Per cent. of total nitrogen, non-albuminoid	Water	Ash	Fat	Nitrogen-free extract	Crude fiber	Albuminoids
INDIANA.																
Head not out	June 8		7.94	1.07	49.93	20.19	10.97	1.75	.18	10.3	70.00	2.38	.59	14.98	8.76	3.28
Before bloom	June 15		7.64	2.27	52.64	29.65	7.40	1.25	.28	22.4	67.50	2.48	.74	17.11	9.64	2.53
In bloom	June 20		7.05	2.18	52.90	32.26	5.52	.88	.00	.0	64.50	2.50	.78	18.81	11.45	1.96
After bloom	July 6		6.63	2.55	52.93	31.32	5.57	.89	.03	3.3	56.30	2.90	1.11	23.57	13.69	1.43
Early seed	July 16		5.93	3.74	60.77	24.70	4.84	.78	.00	.0	53.00	2.80	1.76	28.56	11.61	2.27
NEW HAMPSHIRE.																
Spike invisible			5.19	4.00	57.00	23.46	7.60	1.55	.30	19.4						
Spike visible			4.73	4.22	56.10	23.34	9.61	1.54	.45	29.2						
In bloom			4.57	4.20	57.16	28.28	5.70	.93	.10	11.9						
After bloom			3.88	3.23	58.72	28.92	5.25	.84	.15	17.9						
Early seed			3.2	2.7	62.5	26.03	5.41	.87	.18	20.7						
III.—DACTYLIS GLOMERATA.																
DEPARTMENT GARDEN.																
Panicles not out	Apr. 23	35	10.29	4.13	50.86	18.76	15.97	2.49	1.01	40.6	78.8	2.18	.87	10.78	3.98	3.39
Panicles closed	May 4	55	8.26	3.13	55.04	23.18	10.39	1.63	.00	0.9	79.3	1.71	.64	11.40	4.80	2.15
In full bloom	May 13	87	8.07	3.24	53.76	25.40	9.53	1.53	.16	10.5	77.3	1.63	.74	12.20	5.77	2.16
After bloom	June 1	125	9.01	2.83	52.63	27.26	8.25	1.32	.33	25.0	73.5	2.39	.75	13.05	7.22	2.19
Later growth:																
In bloom	June 18	80	8.64	3.08	50.20	24.67	12.51	1.99	.77	38.7	66.9	2.80	1.32	16.62	8.16	4.14
Late bloom	June 23	75	6.00	3.02	57.34	24.42	8.62	1.38	.42	30.4	60.2	2.39	1.44	22.82	8.72	3.43
Seed nearly ripe	July 1	73	6.73	3.34	57.34	25.09	7.30	1.16	.45	38.8	62.3	2.54	1.26	21.60	9.40	2.75
First year's growth																
Head not out	June 12	28	11.50	6.89	48.06	20.63	12.92	2.07	.15	7.3	70.50	2.36	1.41	9.83	4.23	2.65
Green	July 14		10.52	6.88	46.95	21.64	14.03	2.25	.39	17.3	72.30	2.91	1.90	13.00	6.00	3.89
Yellow	July 15		10.14	5.93	52.37	22.44	9.10	1.46	.18	12.3	74.60	2.58	1.51	13.30	5.70	2.31
	Oct. 25		10.95	6.50	47.98	21.24	13.33	2.14	.54	25.2	68.70	3.43	2.03	15.02	6.65	4.17
IV.—ALOPECURUS PRATENSIS.																
Head just appearing	Apr. 19		9.21	4.09	52.16	18.21	15.73	2.52	.66	38.2	77.1	2.11	1.08	11.94	4.17	3.60
Before bloom	Apr. 19		7.90	4.46	51.66	22.40	13.58	2.17	.53	40.9	76.7	1.84	1.04	12.03	5.22	3.17

	Date	No.															
In bloom	May 1		7.75	3.36	54.70	23.78	10.81	1.73	.09	0.6	60.0	3.10	1.34	21.73	9.51	4.32
After bloom	May 12		8.17	3.39	54.55	25.30	9.6	1.38	.07	5.0	66.6	2.73	1.17	18.15	8.47	2.83
V.—POA PRATENSIS.																
DEPARTMENT GARDEN.																
Set No. 1, grown on good soil:																
Panicle just visible	Apr. 23	20	8.07	4.88	48.74	18.43	10.88	3.18	.48	15.1	76.70	1.68	1.14	11.34	4.30	4.64
Panicle spreading	May 1	30	5.31	4.07	51.32	22.83	16.21	2.68	.30	11.2	70.80	1.61	1.19	14.99	6.67	4.74
In full bloom	May 21	70	8.30	3.90	51.43	23.76	12.01	2.01	.02	1.0	71.90	2.33	1.10	14.43	6.68	3.54
In seed	June 5	70	6.38	4.25	52.54	24.34	12.49	2.00	.37	18.5	53.90	2.81	1.87	21.17	10.74	5.31
Set No. 2, grown on poor soil:																
Panicle closed	Apr. 27	65	6.61	3.92	55.32	21.92	12.23	1.96	.12	6.1	69.00	2.18	.88	17.62	7.90	2.42
In full bloom	May 8		7.02	3.63	50.85	25.46	7.82	1.28	.10	7.8						
Set No. 3, grown on poor soil; wayside:																
After bloom; brown	June 1	65	7.23	3.82	56.19	23.85	8.88	1.42	.25	17.6	55.40	3.22	1.73	23.09	10.64	3.96
In full bloom	May 19	78	7.73	3.41	55.92	23.10	10.44	1.67	.14	8.4	60.20	2.63	1.15	19.69	7.81	3.53
In seed; brown	June 8	75	6.21	3.51	58.58	24.34	7.36	1.18	.15	12.7	54.60	2.81	1.59	26.61	11.05	3.34
QUINCY, ILL.																
Set No. 4:																
Before bloom	May 10		8.42	4.99	45.84	21.87	12.38	3.10	.63	20.3						
In bloom	May 17		7.83	3.77	48.29	24.93	13.09	2.41	.51	21.1						
After bloom	May 27		9.07	3.30	52.51	22.75	12.37	1.97	.35	17.8						
VI.—POA COMPRESSA.																
Poor soil:																
Panicle not out	June 1	14	7.75	5.29	58.04	18.19	10.60	1.71	.10	5.8	67.90	.49	1.70	18.64	5.84	3.43
Panicle well out	June 8	28	6.81	4.41	55.18	21.30	12.30	1.67	.52	20.2	64.70	.64	1.36	17.27	6.67	3.45
In bloom	June 17	30	6.08	4.52	55.14	18.53	12.09	2.03	.45	22.2	70.70	1.74	1.32	17.05	6.43	3.72
After bloom	June 23	30	5.13	3.85	63.80	18.16	8.97	1.43	.35	24.5	51.80	2.47	1.86	30.79	8.75	4.33
VII.—BROMUS UNIOLOIDES.																
Panicle not out	Apr. 23	35	10.05	5.03	48.73	18.34	17.05	2.73	1.06	38.8	80.60	2.07	.97	9.45	3.00	3.31
Panicle closed	May 4	64	8.95	3.44	51.03	22.22	14.36	2.31	.55	23.8	75.40	2.91	.45	12.53	5.47	3.53
In full bloom	May 13	76	9.20	3.94	51.46	22.09	12.63	2.02	.34	16.8	73.40	1.91	.82	10.60	4.67	3.06
After bloom	June 1	70	6.64	2.37	54.79	25.33	10.83	1.74	.35	20.1	67.50	2.17	.77	17.40	8.23	3.12
In seed; brown	June 1	85	8.55	2.10	59.71	19.85	9.79	1.57	.33	21.0	64.70	3.02	.74	21.08	7.01	3.45
VIII.—BROMUS ERECTUS.																
Very young	Apr. 27	35	8.03	3.67	45.84	26.65	16.78	2.52	.43	17.1	85.50	1.25	.53	6.57	3.87	2.28
Before bloom	May 8	60	7.26	3.27	52.01	25.24	12.22	1.95	.24	12.3	74.30	1.80	.84	13.37	6.49	3.14
Early bloom	May 12	64	7.40	3.72	53.38	24.44	11.03	1.76	.00	5.1	72.20	2.00	1.03	14.81	6.81	3.01
After bloom	May 19	73	7.70	2.81	55.19	24.52	8.78	1.41	.34	24.1	63.70	2.79	1.02	20.40	8.90	3.19
	June 1		8.51	2.92	56.32	21.64	8.01	1.38	.40	29.0						

When cut.	Height in centimeters.	Ash.	Fat.	Nitrogen-free extract.	Crude fiber.	Albuminoids.	Total nitrogen.	Non-albuminold nitrogen.	Per cent of nitrogen, non-albuminoid.	Water.	Ash.	Fat.	Nitrogen-free extract.	Crude fiber.	Albuminoids.
IX.—HOLCUS LANATUS.															
Very young .... Apr. 2	..	9.93	4.53	54.48	18.64	12.37	.98	.21	10.6	82.3	1.77	.80	9.61	3.70	2.10
Late bloom .... May 25	72	8.23	3.89	55.53	25.01	7.35	1.30	.60	46.2	50.6	4.07	1.02	27.43	12.5.	3.63
X.—ARRHENATHERUM AVENACEUM.															
In full bloom .... May 25	85	7.93	4.03	54.93	24.35	8.78	1.41	.15	10.6	62.3	2.99	1.52	20.71	9.17	3.31
After bloom .... June 4	60	7.88	4.19	51.76	21.51	14.66	2.35	.00	40.9	74.4	2.02	1.07	13.25	5.51	3.75
XI.—SETARIA GLAUCA.															
Very young .... July 1	50	10.84	2.34	48.12	21.68	17.02	2.72	1.00	36.8	74.2	2.80	.60	12.42	5.59	4.30
Early flowering .... July 24	80	7.25	2.66	55.28	25.75	9.04	1.44	.41	28.5	68.4	2.29	.81	17.47	8.14	2.86
XII.—ANTHOXANTHUM ODORATUM.															
Very young .... May 1	15	6.39	4.27	61.58	17.17	10.50	1.70	.06	3.5	76.9	1.47	.99	14.22	3.97	2.45
In full bloom .... May 1	40	7.09	4.36	59.45	20.53	9.47	1.53	.15	9.3	78.8	1.50	.71	12.62	4.37	2.00
After bloom .... June 19	45	7.27	4.86	53.40	21.17	13.30	2.13	.51	23.9	69.9	2.20	1.46	16.07	6.37	4.00
After blooming .... July 19	55	5.79	4.08	58.02	25.00	7.11	1.14	.33	30.7	53.4	2.70	1.00	27.04	11.65	3.31
XIII.—FESTUCA OVINA.															
Very young .... Apr. 27	25	6.47	4.31	54.00	20.31	14.91	2.38	.12	5.0	70.0	1.94	1.29	16.21	6.09	4.47
Before bloom .... May 8	36	5.41	3.61	57.13	23.10	8.75	1.40	.06	4.3	65.5	1.87	1.25	19.76	8.69	3.03
Do .... May 12	45	6.00	3.43	55.44	23.65	9.48	1.53	.16	10.5	67.0	1.98	1.13	18.29	8.47	3.13
In bloom .... May 21	40	5.60	2.51	58.20	23.79	8.90	1.58	.27	17.1	53.7	2.59	1.16	26.95	11.02	4.58
After bloom .... June 1	47	6.57	3.07	57.09	23.96	9.31	1.49	.27	18.1	53.9	3.03	1.41	28.32	11.05	4.29
XIV.—LOLIUM PERENNE VAR. ITALICUM.															
Heads invisible .... Apr. 27	55	13.28	4.69	42.04	18.15	21.84	3.46	.07	19.8	83.3	2.35	.85	7.45	3.22	3.83
Heads just out .... May 21	75	11.39	3.81	48.74	21.75	14.31	2.29	.39	17.0	82.7	1.97	.06	8.43	3.76	2.48
In full bloom .... May 26	90	11.02	2.32	51.73	20.44	14.49	2.32	.18	7.8	78.0	2.42	.51	11.39	4.49	3.19
After bloom .... June 4	92	8.76	3.08	53.81	21.86	11.59	1.85	.43	23.2	71.5	2.50	1.13	15.34	6.23	3.30
First year's growth: Head not out .... June 2	22	13.21	6.91	45.55	15.50	18.80	3.01	.60	19.9	84.00	2.12	1.16	7.29	2.48	3.01
Do .... June 12	31	12.70	6.36	49.69	16.99	14.26	2.28	.45	19.7	82.70	2.20	1.10	8.59	2.91	2.47

Do	June 10	38	14.06	6.18	45.07	17.84	16.85	2.09	.60	24.6	85.30	2.49	1.09	7.98	3.16	2.98
Do	July 10		13.87	6.53	44.50	20.65	14.45	2.31	.69	25.5	78.90	2.82	1.38	8.39	4.36	3.05
Do	Oct. 25	25	10.87	5.31	47.62	22.40	13.60	2.17	.49	22.6	71.80	3.09	1.51	13.58	6.36	3.86

XV.—LOLIUM PERENNE.

Head invisible	May 1	35	8.68	3.58	57.70	18.39	11.67	1.87	.28	15.0	78.6	1.85	.76	12.35	3.04	2.50
Do	Mar 4	28	9.48	4.34	55.08	18.00	13.10	2.09	.39	18.7	82.4	1.67	.76	9.69	3.17	2.31
Head well out	May 4	30	7.90	3.64	56.75	20.55	11.10	1.78	.31	18.5	74.0	2.07	.94	14.71	5.34	2.89
Before bloom	May 12	55	8.40	3.75	54.93	23.93	8.99	1.43	.09	6.3	76.4	1.98	.89	12.96	5.65	2.12
After bloom	June 1	52	7.50	2.64	56.84	25.32	7.60	1.21			63.1	2.77	.97	20.97	9.38	2.81

The preceding analyses furnish the data from which is derived the general conclusion that as a grass grows older its content of water decreases, ash decreases, fat decreases, albuminoids decrease, carbohydrates increase, crude fiber increases, non albuminoids decrease till bloom or slightly after, when it is at its lowest, and then increases again during the formation of the seed.

There are exceptions to these rules, but for the large majority of species under ordinary conditions of environment they hold good.

There are almost no exceptions to the fact that the water decreases in the maturer specimens; that is to say, the plant gradually dries up and becomes less succulent. The ash is very dependent on locality and surroundings, and as in the analyses which are here published it includes whatever soil there may be mechanically adherent to the blade or stalk as collected, it sometimes shows irregularities from one period to another.

The albuminoids decrease in amount with great regularity, the few cases where an increase appears being due to the fact that the specimens were probably grown under varying conditions.

The fiber sometimes decreases, as in *Bromus erectus*, but the change in that direction is never large. .

The non-albuminoid constituents, however, are often quite the reverse of constant in their manner of appearance and disappearance, and show themselves to be largely or more affected by environment than any other constituent. In *Agrostis vulgaris* they continue to decrease after bloom, and in *Anthoxanthum odoratum* and *Festuca ovina* they increase steadily from early growth to maturity. The relative amount present in the same species from different localities is extremely variable, as may be seen in the analyses of *Phleum pratense*, where specimens from Indiana contain almost no non-albuminoid nitrogen, while those from the Dis. trict of Columbia and elsewhere are well supplied. The specimens from poorer soil having the smallest amount in some cases and the largest in others, the fact can hardly be due directly and entirely to the lack of cultivation, but as the *averages* show that the best grasses contain the least non-albuminoids it is plain that it is dependent on the sources of nitrogen and the supply furnished the plant. The usual changes in the non-albuminoids seem to point to the possibility that they increase at the time of the formation of the seed in the act of transferring to the seed, as amides, the nitrogen of the plant.

THE BEST PERIOD OF GROWTH AT WHICH TO CUT FOR HAY.

Although largely a matter of opinion, it would seem from the forego. ing results that the time of bloom or very little later is the fittest for cutting grasses to be cured as hay. The amount of water has diminished relatively, and there is a proportionately larger amount of nutriment in the material cut, and the weight of the latter will be at its highest point economically considered. Later on, the amount of fiber becomes too prominent, the stalk grows hard, arid, indigestible, and the albuminoids decrease, while the dry seeds are readily detached from their glumes and lost with their store of nitrogen.

For different species, however, different times are undoubtedly suitable, and experience must be added to our chemical knowledge to enable a rational decision to be arrived at.

THE COMPOSITION OF THE ASH.

The ash of many foreign varieties of grasses have been analyzed and the results collected and published by Wolff. Of American growth the ash of only a number of the wild grasses collected in 1878 have been examined. The results are here given:

Ash analyses—Grasses.

Name.	Phosphoric acid, P_2O_5	Sulphuric acid, SO_3	Silica, SiO_2	Chlorine, Cl	Calcium oxide, CaO	Magnesium oxide, MgO	P o t. oxide, K_2O	Potassium, K	S o d. oxide, Na_2O	Sodium, Na
Hierochloa borealis (Vanilla Grass)	7.42	2.55	42.73	4.49	3.07	2.54	31.51	4 54	.....	.25
Eleusine Indica (Wire Grass)...........	2.69	4.24	47.56	10.09	10.27	4.10	10.27	0.52	.....	1 26
Eleusine Indica (Wire Grass)...........	9.68	5.70	24.61	6.71	13.65	7.38	24.79	7.39	.....	.....
Eleusine Indica (Wire Grass)	9.84	8.55	16.25	0.61	11.10	5.57	30.08	4.55	.....	3.55
Uniola latifolia (Fescue Grass).........	4.92	2.02	66.87	4.71	7.15	3.02	5.52	5.19	.....	.....
Cynodon Dactylon (Bermuda Grass) ...	6.20	9.37	30.29	6.05	13.44	5.00	22.09	6.66	.....	.....
Cynodon Dactylon (Bermuda Grass)....	5.09	11.31	30.27	9.46	7.09	2.96	22.89	9.61	.....	.42
Sporobolus Indicus (Smut Grass)	6.02	4.60	27.36	11.03	2.64	2.66	33.53	12.16	.....	.....
Andropogon Virginicus (Broom Grass)..	2.97	2.80	58.33	6.37	6.76	1.83	13.93	7.01	.....	.....
Andropogon scoparius	1.33	trace.	64.62	15.65	2.12	.58	.....	15.70	.....	.....
Poa pratensis (Kentucky Blue Grass) ..	9.88	4.76	30.25	6.30	4.81	3.23	33.81	6.95	.....	.....
Poa serotina (Fowl Meadow Grass) ...	10.80	3.35	37.10	3.80	6.70	2.92	31.71	2.79	.....	.83
Dactyloctenium Egypt.(Egyptian Grass)	8 37	4.42	34.17	6.76	20.67	6.91	21.20	7.50	.....	.....
Panicum sanguinale (Crab Grass).......	6.40	4.02	30.93	6.04	4.40	7.98	33.56	6.67	.....	.....
Panicum maximum (True Guinea Grass)	4.37	2.51	16.51	7.77	10.18	14.16	35.93	8.57	.....	.....
Panicum obtusum	5.18	6.71	48.60	4.20	5.91	3.13	21.65	4.62	.....	.....
Panicum virgatum (Tall Panic Grass)	5.50	3.56	51.17	4.93	7.87	3.63	18.76	3.36	.....	1.22
Panicum sp.?	4.37	5.29	45.10	4.06	7.30	7.98	22.53	1.54	.....	1.74
Panicum Tezanum (Texas Millet......	8.48	4.63	34.31	6.55	7.39	4.57	27.95	4.54	.....	1.58
Panicum Crus-galli (Barnyard Grass) ..	4.27	3.69	42.18	11.48	7.23	5.52	13.26	12.00	.....	.37
Panicum filiforme (Slender Crab Grass)	8.37	4.84	40.36	4.11	4.69	5.18	12.96	13.41	.....	.....
Sorghum halepense (Johnson Grass) ...	10.44	2.96	22.21	4.58	12.87	6.73	35.72	3.68	.....	.81
Chrysopogon nutans (Indian Grass) ...	2.35	2.13	61.56	6.11	2.92	1.36	16.84	6.74	.....	.....
Muhlenbergia diffusa (Drop Seed).......	6.65	3.39	30.98	8.21	11.95	4.30	17.32	0.78	.....	1.33
Bromus unioloides (Schrader's Grass)...	8.79	5.61	4.84	16.84	4.43	4.64	37.20	16.33	.....	1.27
Bromus carinatus (Broom Grass)......	9.29	3.94	38.33	3.30	6.19	2.19	31.61	.....	2.17	2.98
Agrostis exarata (Browntop)	8.01	1.93	34.63	3.60	5.61	3.34	38.41	3.97	.....	.....
Paspalum læve (Water Grass)	6.18	5.64	44.65	1.73	9 36	5 26	25.44	.....	.60	1.12
Setaria setosa (Bristly Fox-tail)	3.24	3.51	42.59	3.81	6.21	1.56	39.33	.....	1.18	2.47
Leptochloa mucronata (Feather Grass)..	6.46	3.31	55.92	2.80	5.04	2.66	20.21	1.81	.....	.80
Tripsacum dactyloides (Gama Grass)...	2.52	3.09	37.84	13.08	1.64	1.07	29.06	6.30	.....	4.77
Tricupis sesterioides (Tall Redtop)	1.58	4.04	37.52	7.39	2.32	.53	38.49	8.13	.....	.....

CONCLUSION.

The work which has been collected in the previous pages extended over several years, from 1878 to 1883. It was inaugurated by Dr. Peter Collier, as chemist to this Department, and the laboratory work for the first year was in the hands of Mr. Henry B. Parsons, Mr. Charles Wellington, and myself. The remainder of the work has been under my own supervision, and has been almost entirely carried out by Mr. Miles Fuller and myself. It is hoped that the collection and re-arrangement of the results will give them an increased value.

GLOSSARY OF TERMS USED IN DESCRIBING GRASSES.

Abrupt. Terminating suddenly.
Acuminate. Extended into a tapering point.
Acute. Sharp-pointed.
Alternate. Situated regularly one above the other on opposite sides.
Annual. Living but one season.
Anther. The organ containing the pollen or flower-dust.
Apex. The top or extreme end of any part.
Appressed. Pressed together, not spreading.
Aristate. Having an awn or beard.
Articulated. Connected by a joint or joints.
Ascending. Rising obliquely from the ground.
Awl-shaped. Gradually narrowed to a fine point like an awl.
Awn. A bristle-like organ proceeding from the glumes.
Axis. The central stem of a panicle, spike, or spikelet on which the flowers are disposed.
Beard. A long slender hair or awn.
Biennial. Living through two seasons.
Bifid. Divided into two portions at the apex.
Bisexual. Having both stamens and pistils.
Blade. The expanded portion of a leaf.
Boat-shaped. Folded together in the form of a boat; convex outwardly and concave on the inside.
Branch. A division of the stem or panicle.
Branchlet. A secondary division of the branch.
Bristle. Short, stiff hairs.
Bulbous. Thickened like a bulb.
Cæspitose. Growing in bunches or tufts.
Capillary. Hair-like, very slender.
Carinate. Keeled, having a prominent ridge in the center.
Cartilaginous. Firm and tough like cartilage.
Cauline. Belonging to the culm or stem.
Chaff. The dried glumes and palets of grasses.
Chartaceous. The texture resembling paper or parchment in thickness.
Ciliate. Having the margin or nerves fringed with hairs.
Compressed. Flattened laterally.
Contorted. Twisted.
Convolute. Rolled together inward from the margins.
Coriaceous. Of a leathery consistence.
Cornaceous. Of a horn-like consistence.
Culm. The stalk or stem of grasses.
Cuspidate. Ending in a sharp, stiff point.
Decumbent. Reclining on the ground, but rising at the top.
Dichotomous. Branching in twos, forking by pairs.
Digitate. Dividing from a common point.

141

Diœcious. Having the stamens and pistils on separate plants, the staminate flowers on one and pistillate flowers on another.

Diverging. Widely spreading.

Dorsal. Belonging to or growing from the back.

Emarginate. Having a notch at the end.

Entire. Without notches or divisions.

Equal. Alike in length.

Exserted. Protruded, extended beyond, standing out.

Fertile. Having perfect pistils; producing fruit.

Fibrous. Having thread-like divisions.

Filament. The stalk or support of the anther.

Filiform. Thread-like.

Flexuous. Bending freely.

Floret. The flowers of grasses are sometimes called florets.

Foliaceous. Resembling a leaf.

Fusiform. Spindle-shaped, largest in the middle and tapering to both ends.

Geniculate. Bent abruptly, like a knee.

Genus. A group of species having a general agreement in structure.

Glabrous. Smooth, without hairs or roughness.

Glaucous. Having a grayish green color.

Glomerate. Clustered in small roundish heads.

Glumes. The chaff-like leaves forming a part of the flowers.

Herbaceous. Herb-like, not woody.

Hirsute. Pubescent, with rather stiff and coarse hairs.

Hyaline. Thin and transparent.

Imbricate. Closely overlapping each other, as frequently the flowers of a spikelet.

Indigenous. Growing naturally; not brought from some other country.

Inferior. Lower in position.

Inserted. Growing out of, or upon another.

Internode. The space between two nodes or joints.

Involute. Rolled together inwards.

Joints. Thickenings in the stem where the leaves originate; separable parts of an axis; point of issue of the branches of a panicle.

Keel. An elevated longitudinal ridge, in the middle of a leaf, glume, or palet; resembling the keel of a boat.

Lamina. The free or expanded portion of a leaf, as distinguished from the petiole or the sheath; the blade of a leaf.

Lanceolate. Tapering gradually to the apex, like a lancet.

Lateral. At or from the side.

Ligule. A tongue-like appendage at the upper part of the sheath of a leaf.

Line. The twelfth part of an inch.

Linear. Long and narrow, with parellel sides.

Lobe. Some division of a glume.

Male flower. A flower that has stamens, but without pistil.

Membranaceous. Thin like a membrane, generally somewhat translucent.

Monœcious. With stamens and pistils in different flowers on the same plant.

Midrib. The central and principal nerve of a leaf or glume.

Mucronate. Abruptly tipped with a short awn or bristle.

Nerves. The ribs or veins of a leaf, or leaf-like organ.

Neutral. Having neither stamens nor pistils.

Nodes. Knots or thickened portions in the culms.

Oblong. Longer than wide, with nearly parellel sides.

Obovate. Egg-shaped, with the wider end uppermost.

Obtuse. Blunt or rounded at the apex.

Oval. Broadly elliptical, approaching a rounded form.

Orary. That part of the pistil which contains the seed.

Ovate. Egg-shaped.

Palet or palea. The inner scale or chaff of the proper flower, placed nearly opposite and a little higher on the axis than the flowering glume.

Panicle. A branched and subdivided stem bearing the flowers.

Pedicel. A small branchlet supporting a spikelet.

Peduncle. The main stem or stalk of a flower-spike.

Perennial. Living more than two years, or indefinitely.

Perfect. Having both stamens and pistils.

Petiole. The stem of a leaf.

Pistil. The central or female organ of a fertile flower.

Pistillate. Having only pistils without stamens.

Plumose. Feather-like.

Pollen. The powder contained in the anthers.

Pubescent. Clothed with short and soft hairs.

Radical leaves. Those growing from the base of the culm.

Revolute. Ro ling backwards or outwards.

Rhachis. The axis or stem on which the flowers of a spikelet are arranged; also the common axis of a close spike or of a panicle.

Rhizoma or Root-stock. A horizontal underground stem.

Ribs. Prominent nerves of the leaves or glumes.

Rugose. Wrinkled or furrowed.

Scabrous. Rough, with small points or short stiff hairs.

Scarious. Dry and thin, and generally transparent.

Sericeous. Covered with soft, silky hairs.

Serrate. Having teeth on the margin, pointed towards the apex.

Serrulate. Finely-toothed.

Sessile. Without a foot-stalk or pedicel.

Setaceous. Like a bristle.

Sheath. That part of the leaf which embraces the culm or stalk.

Spike. A collection of sessile or nearly sessile flowers on a close, narrow axis.

Spikelet. A flower or cluster of flowers having one pair of outer glumes.

Stamen. The male organs of a flower, including the anther and filament.

Staminate. Having only stamens.

Sterile. Imperfect flowers not producing seed.

Strict. Erect and close.

Stoloniferous. Sending off offshoots or runners from the base.

Strigose. Having spreading, bristly hairs.

Style. That portion of the pistil bearing stigmas or a stigmatic surface; in grasses often branching.

Subulate. Stiff and linear, shaped like an awl.

Succulent. Fleshy and juicy.

Truncate. Abruptly cut off at the apex.

Unequal. Not of equal length.

Verticillate. Arranged in a whorl or whorls.

Villous. Velvety, clothed thickly with long, soft hairs.

Whorl. A number of leaves or branches arranged around a stem on the same plane.

Woolly. Clothed with long and matted hairs.

INDEX OF PLATES.

	Plate.
Agropyrum glaucum	67
repens	88
Agrostis	48
exarata	49
exarata var. Pacifica	106
vulgaris	48
Alfalfa	96
Alfilaria	101
Alopecurus pratensis	35
Andropogon furcatus	29
macrourus	28
scoparius	27
Virginicus	26
Anthoxanthum odoratum	34
Aristida purpurea	35
Arrhenatherum avenaceum	58
Avena fatua	57
Barnyard grass	14
Beckmannia crucaeformis	8
Bermuda grass	79
Blue grass, English	74
Kentucky	75
mountain	78
Texas	73
Bluejoint	29, 51
Blue stem, Colorado	87
Bouteloua oligostachya	62
racemosa	63
Bromus secalinus	84
unioloides	85
Broom sedge	27
Buchloë dactyloides	66
Buffalo grass	66
Bunch grass	83
Cactus	99
Calamagrostis Canadensis	51
longifolia	53
sylvatica	52
Canary grass, American	33
Cheat	84
Chess	84
Chloris alba	108
Chrysopogon nutans	30
Cinna arundinacea	50
Clover, Alsike	92
buffalo	94
bur	97

3594 GR——10   145

	Plate.
Clover, Japan	98
Mexican	102
Cord grass	60
Couch grass	88
Crab grass	15
Crowfoot	65
Cut grass	24
Cynodon Dactylon	59
Dactylis glomerata	72
Deschampsia caespitosa	107
Diplachne dubia	109
Distichlis maritima	71
Eleusine Aegyptiaca	65
Indica	64
Elymus Canadensis	89
condensatus	90
Virginicus	91
Eragrostis major	70
Erodium cicutarium	101
Esparsette	95
Euchlaena luxurians	22
Eurotia lanata	100
Fescue, sheep's	82
tall meadow	81
Festuca elatior	81
ovina	82
scabrella	83
Fowl meadow grass	76
Foxtail, meadow	35
Gama grass	21
General illustrations of grasses:	
Dissections of flowers	4
Inflorescence	3
Roots and rhizomes	1
Sheaths, ligules, and blades	2
Glyceria arundinacea	79
Canadensis	114
nervata	80
Grama grass	62
black	62
tall	63
Guinea grass	9
Herd's grass	45
Hilaria Jamesii	25
Holcus lanatus	54
Hungarian grass	19
Johnson grass	31
June grass	75
Kœleria cristata	69
Leersia Virginica	24
Lolium perenne	86
Lespedeza striata	98
Lucerne	96
Medicago denticulata	97
sativa	96

	Plate.
Melica bulbosa	111
diffusa	110
imperfecta	112
Millet grass, Indian	13
Millium effusum	41
Muhlenbergia comata	104
diffusa	41
glomerata	42
Mexicana	43
sylvatica	44
Nimble Will	41
Oat grass, tall meadow	58
wild	30
Onobrychis sativa	95
Opuntia Engelmanni	99
Orchard grass	72
Oryzopsis cuspidata	39
Panicum agrostoides	17
anceps	18
barbinode	12
Crus-galli	14
gibbum	103
maximum	9
miliaceum	13
proliferum	11
sanguinale	15
Texanum	10
virgatum	16
Paspalum platycaule	6
Para grass	12
Phalaris arundinacea	32
intermedia	32
Phleum pratense	45
Phragmites communis	68
Pigeon grass	20
Pin grass	101
Poa andina	78
arachnifera	73
compressa	74
pratensis	75
serotina	76
tenuifolia	77
Prickly pear	99
Quack grass	88
Reed grass	68
Redtop	48
western	49
Relief grass, Stewart's	33
Richardsonia scabra	102
Rye grass, perennial	86
Sainfoin	95
Salt grass	71
Setaria glauca	20
Italica	19
Smut grass	47

148

	Plate.
Sorghum halepense	31
Spartina cynosuroides	60
juncea	61
Sporobolus airoides	105
cryptandrus	46
Indicus	47
Stipa spartea	38
viridula	37
Sweet vernal grass	34
Switch grass	16
Teosinte	22
Timothy	45
Trifolium hybridum	92
incarnatum	94
stoloniferum	94
Trisetum palustre	55
subspicatum	56
Triodia seslerioides	67
Tripsacum dactyloides	21
Uniola latifolia	113
Velvet grass	54
White grass	24
Wild oats	57
Wild rice	23
Wild rye	89, 91
Winter fat	100
Wood grass	30
Zizania aquatica	23

O

PLATE 1.

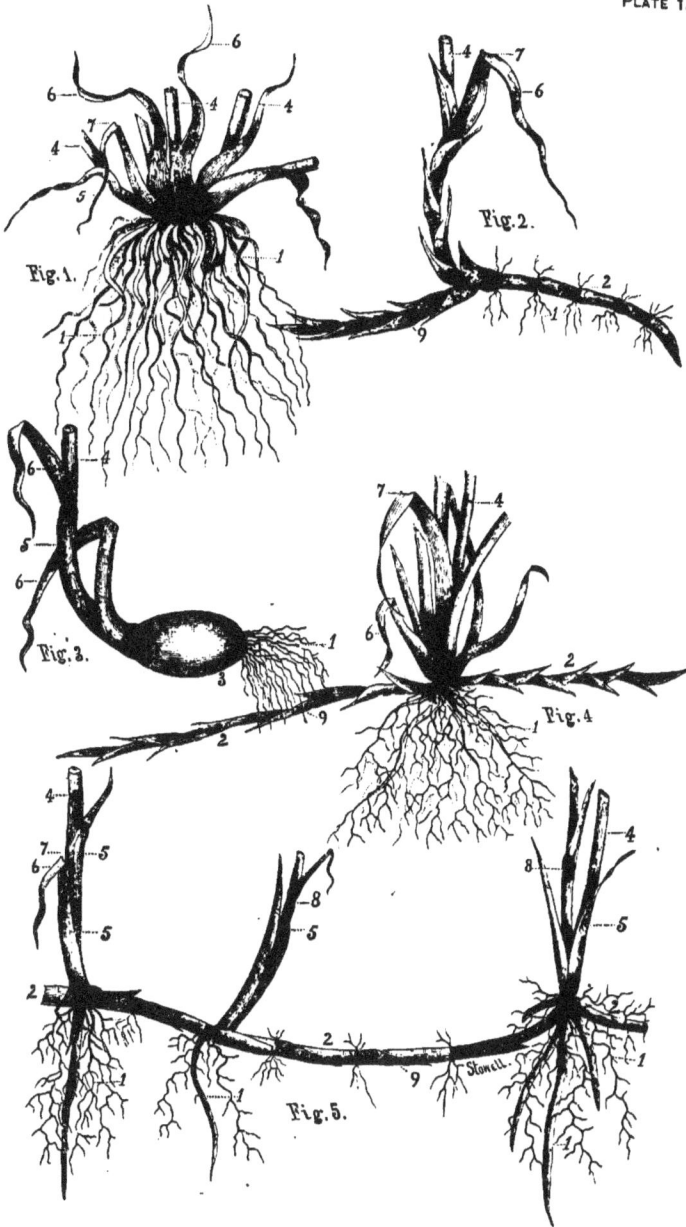

ROOTS AND RHIZOMES OF GRASSES.

PLATE 2.

SHEATHS, LIGULES, AND BLADES OF GRASSES.

PLATE 3.

INFLORESCENCE OF GRASSES.

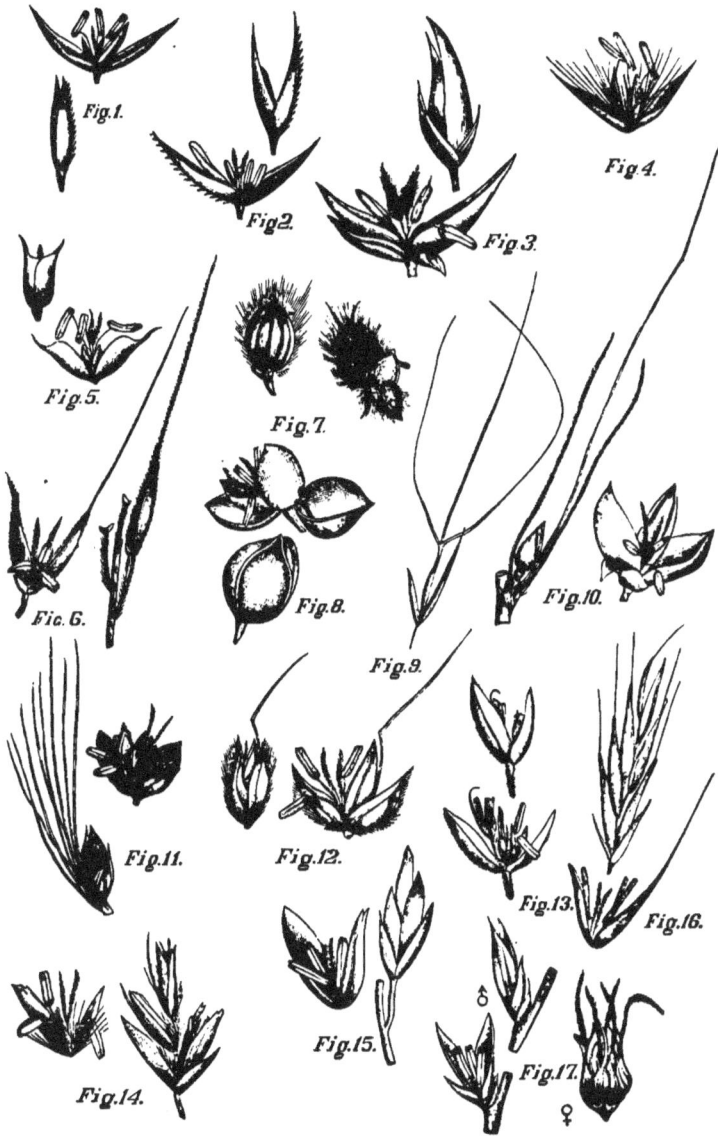

Fig.1.

Fig.2.

Fig.3.

Fig.4.

Fig.5.

Fig.7.

Fig.6.

Fig.8.

Fig.9.

Fig.10.

Fig.11.

Fig.12.

Fig.13.

Fig.14.

Fig.15.

Fig.16.

Fig.17.

DISSECTIONS OF GRASS FLOWERS.

PLATE 5.

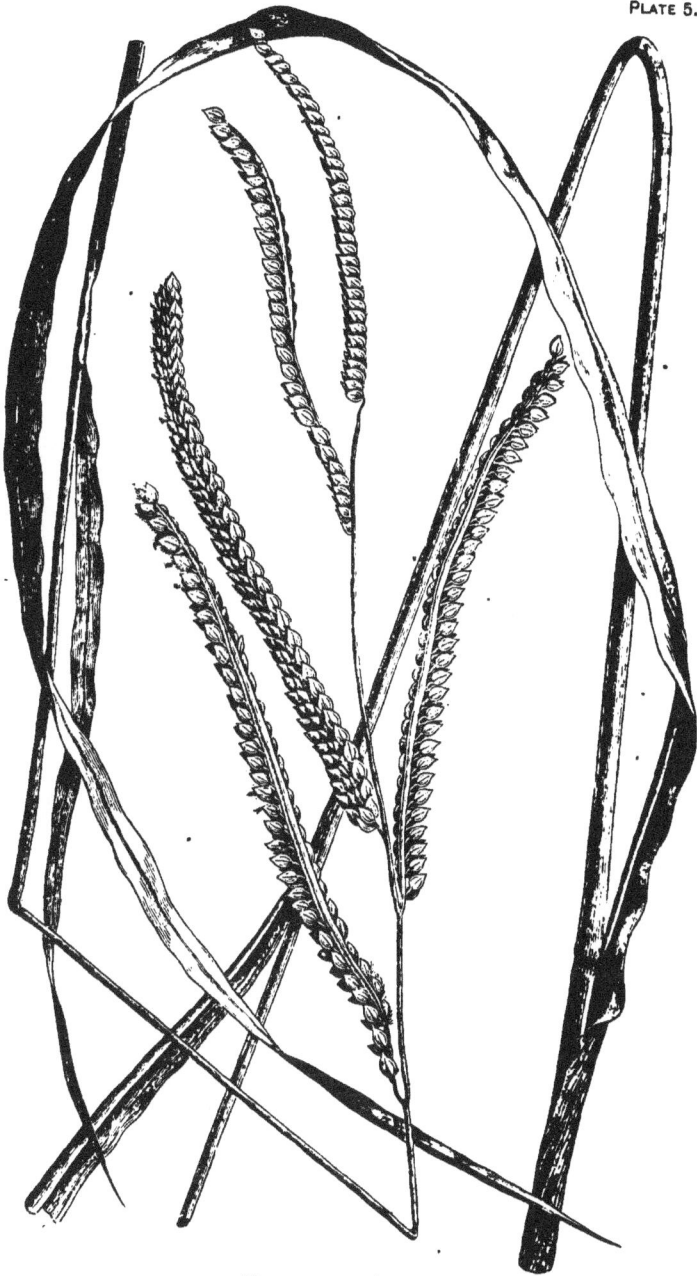

PASPALUM DILATATUM.

W.R.SCHELL.del.

a

b

W.B.Scholl.del.

PASPALUM DISTICHUM.

PLATE 8.

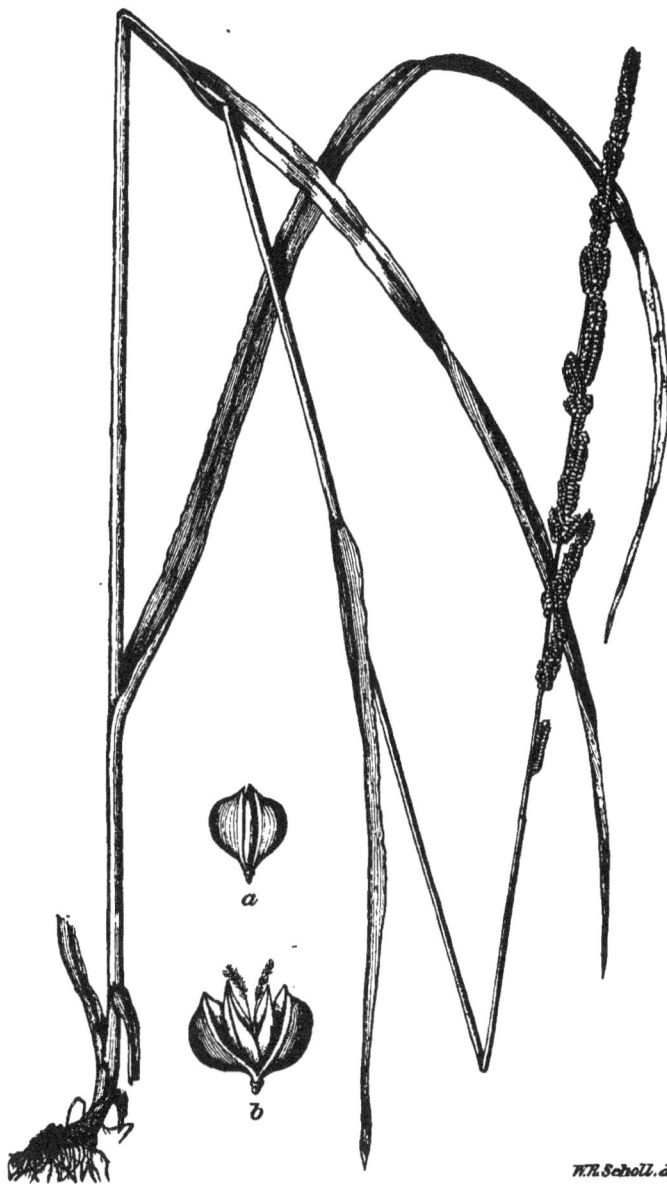

a

b

W.R.Scholl. del.,

BECKMANNIA ERUCÆFORMIS.

PANICUM MAXIMUM, Guinea grass.

Marx del.

PANICUM TEXANUM, Texas blue grass.

PLATE 11

PANICUM BARBINODE, Para grass.

PLATE 13.

W.B.S.

PANICUM MILIACEUM, Indian millet grass.

PLATE 14.

PANICUM CRUS-GALLI, Barn-yard grass.

PLATE 15.

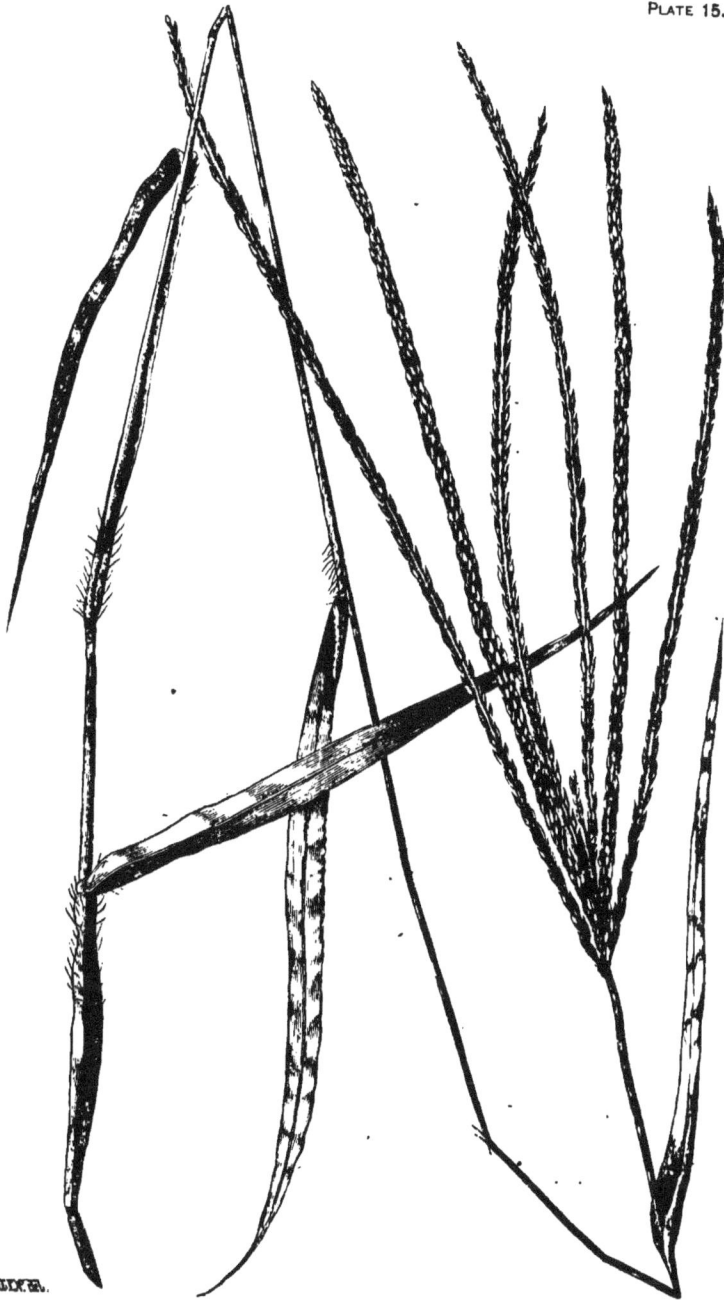

PANICUM SANGUINALE, Crab grass.

PLATE 16.

PANICUM VIRGATUM, Switch grass.

PANICUM AGROSTOIDES.

PLATE 18.

PANICUM ANCEPS.

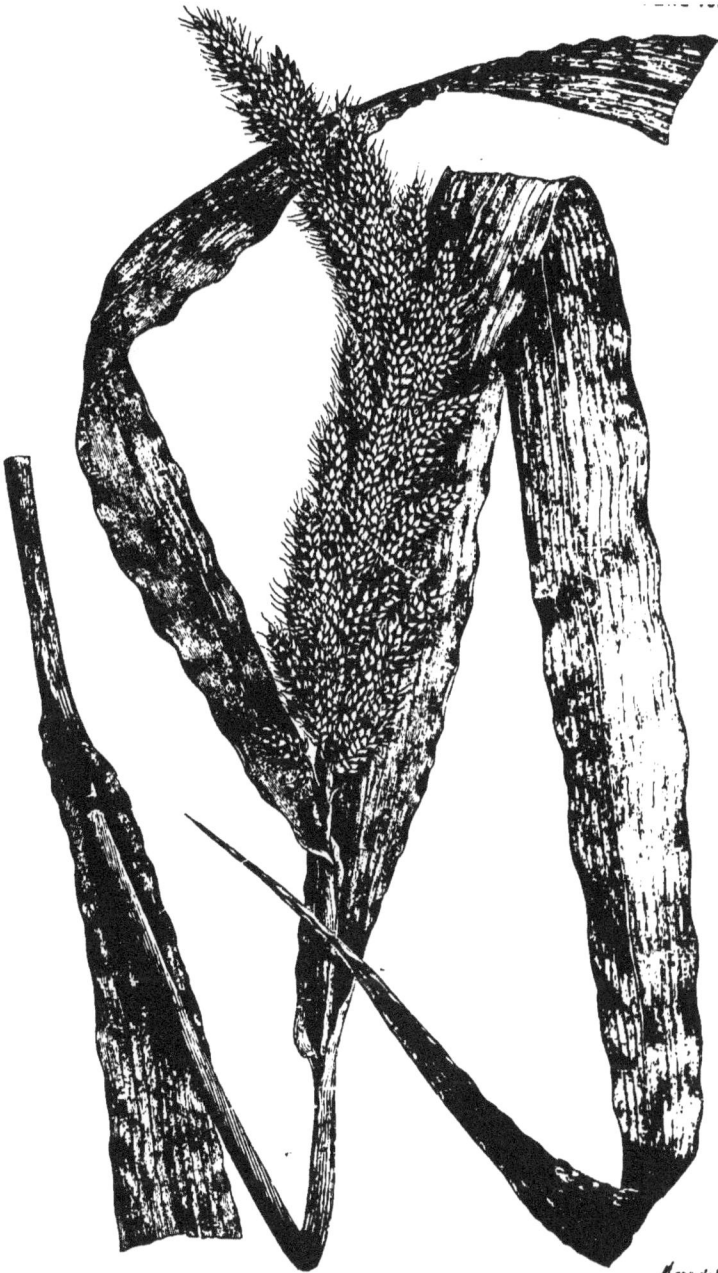

SETARIA ITALICA, Hungarian grass.

PLATE 20.

MARX.—DEL.

SETARIA GLAUCA, Pigeon grass.

PLATE 21.

TRIPSACUM DACTYLOIDES, Gama grass.

PLATE 22.

1

2

3

R.aing

PLATE 23.

ZIZANIA AQUATICA, Wild rice.

PLATE 24.

a

LEERSIA VIRGINICA, White grass, Cut grass.

PLATE 25.

a

b

.

PLATE 27.

ANDROPOGON SCOPARIUS, Broom sedge.

PLATE 28.

MARX.DEL.

NICHOLS.

ANDROPOGON MACROURUS.

PLATE 29.

ANDROPOGON FURCATUS, Bluejoint.

PLATE 30.

CHRYSOPOGON NUTANS, Wild oat grass, Wood grass.

PLATE 31.

SORGHUM HALEPENSE, Johnson grass.

PLATE 32.

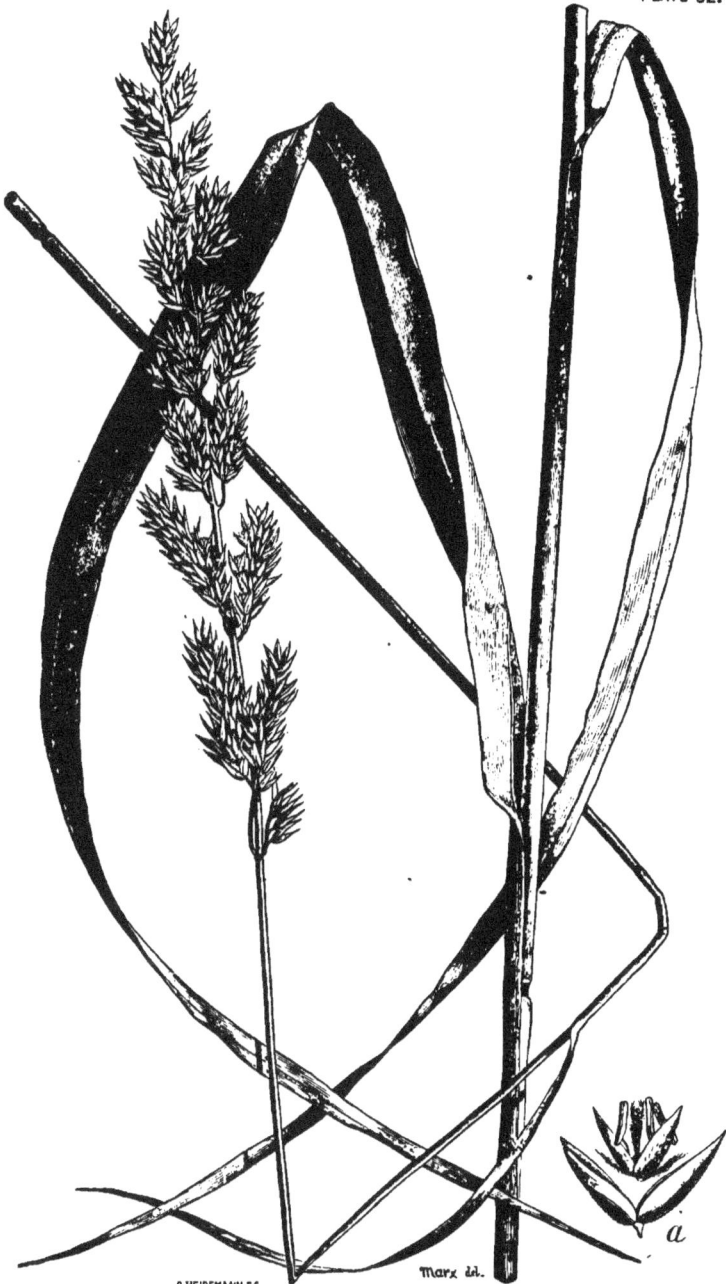

O.HEIDEMANN SC.   Marx del.

PHALARIS ARUNDINACEA.

PLATE 33.

H.H.NICHOLS. SC

PHALARIS INTERMEDIA, Stewart's relief grass, American Canary grass.

PLATE 34.

MARX.DEL.

ANTHOXANTHUM ODORATUM, Sweet vernal grass.

PLATE 35.

S.H.NICHOLS-ENG.

MARX-DEL.

ALOPECURUS PRATENSIS, Meadow foxtail.

PLATE 36.

H.H.NICHOLS-ENG.

MARX.DEL.

ARISTIDA PURPUREA, Awned bunch grass.

PLATE 37.

MARX.Del.

STIPA VIRIDULA.

PLATE 38.

PLATE 39.

ORYZOPSIS CUSPIDATA, Indian millet grass.

PLATE 40.

MILIUM EFFUSUM.

MUHLENBERGIA DIFFUSA, Nimble Will. '

MUHLENBERGIA GLOMERATA.

.

MUHLENBERGIA MEXICANA.

PLATE 44.

F.H.N Sc.

MARX.DE

MUHLENBERGIA SYLVATICA.

PLATE 45.

PHLEUM PRATENSE. Timothy.

PLATE 46.

Marr del.

SPOROBOLUS CRYPTANDRUS.

PLATE 47.

SPOROBOLUS INDICUS, Smut grass.

PLATE 46.

H.H M.

MARX DEL.

AGROSTIS VULGARIS, Red top.

PLATE 49.

AGROSTIS EXARATA, Western red top.

PLATE 50.

CINNA ARUNDINACEA.

PLATE 51.

NICHOLS—ENG.

MARX—DEL.

CALAMAGROSTIS (DEYEUXIA) CANADENSIS. Bluejoint.

PLATE 52.

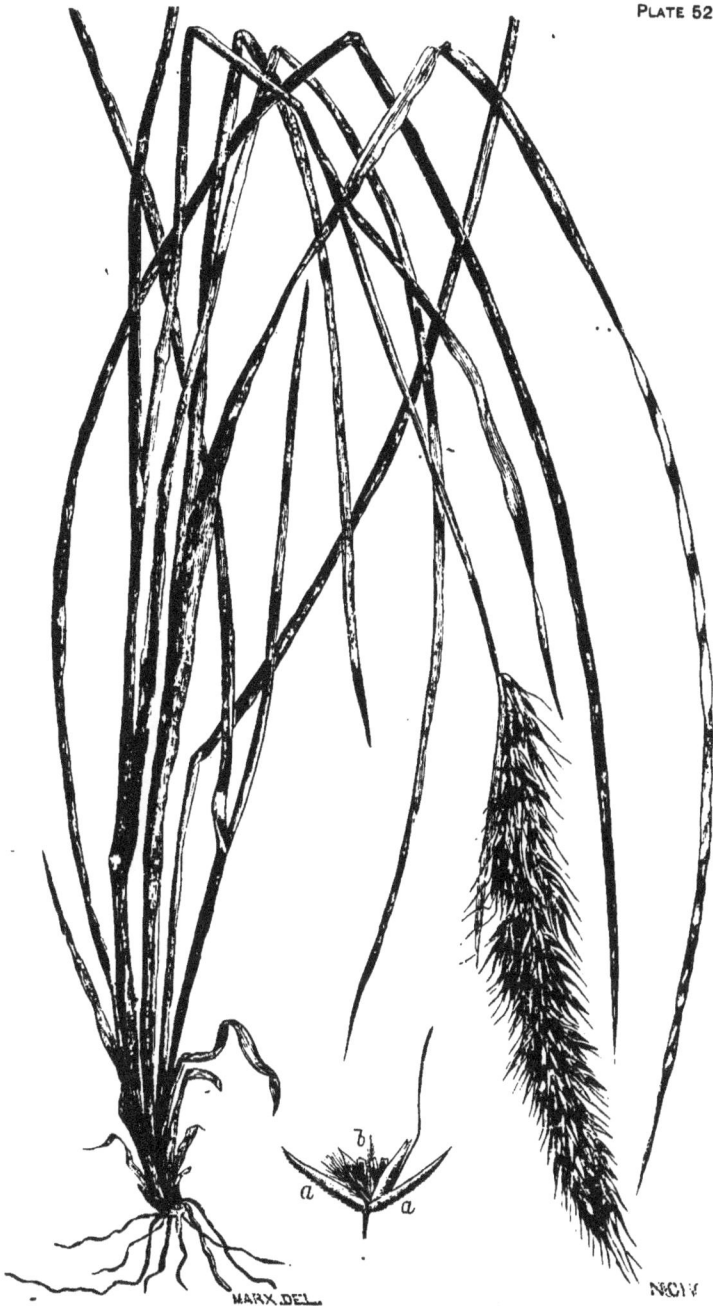

CALAMAGROSTIS (DEYEUXIA) SYLVATICA.

PLATE 54.

'.H.Nichols.Sc   Hootz del.

HOLCUS LANATUS, Velvet grass.

PLATE 55.

Marx del.

TRISETUM PALUSTRE.

PLATE 56.

H.H.NICHOLS.

mapz del

TRISETUM SUBSPICATUM.

PLATE 57.

AVENA FATUA, Wild oats.

PLATE 58.

ARRHENATHERUM AVENACEUM, Tall meadow oat grass.

Cynodon Dactylon, Bermuda grass.

SPARTINA CYNOSUROIDES, Cord grass.

PLATE 61.

SPARTINA JUNCEA.

PLATE 62.

BOUTELOUA OLIGOSTACHYA, Gramma grass, Black gramma.

BOUTELOUA RACEMOSA, Tall gramma grass.

PLATE 64

ELEUSINE INDICA.

PLATE 65.

ELEUSINE ÆGYPTIACA, Crowfoot.

PLATE 66.

NICHOLS

MARX–DEL.

BUCHLOË DACTYLOIDES, Buffalo grass.

TRIODIA SESLERIOIDES.

PLATE 68.

Mary del.

PHRAGMITES COMMUNIS, Reed grass.

PLATE 69.

T.TAYLOR.DEL.

H.H.NICHOLS.

KŒLERIA CRISTATA.

PLATE 70.

ERAGROSTIS MAJOR.

DISTICHLIS MARITIMA, Salt grass.

PLATE 72.

ARX.DEL.

DACTYLIS GLOMERATA, Orchard grass.

PLATE 73.

POA ARACHNIFERA, Texas blue grass.

H.H.N

PLATE 74.

POA COMPRESSA, English blue grass.

PLATE 75.

Marx del.

POA PRATENSIS, Kentucky blue grass, June grass.

PLATE 76.

POA SEROTINA, Fowl meadow grass.

PLATE 77.

POA TENUIFOLIA.

MARX.DEL.

POA ANDINA, Mountain blue grass.

PLATE 71

GLYCERIA ARUNDINACEA.

PLATE 80.

GLYCERIA NERVATA.

PLATE 81.

CX.DEL

NICHOLS

PLATE 82.

H.H.NICHOLS.EN

PLATE 83.

FESTUCA SCABRELLA, Rough-leaved fescue.

PLATE 84.

BROMUS SECALINUS, Chess.

MARX.DEL.

PLATE 85.

BROMUS UNIOLOIDES (half size).

PLATE 86.

LOLIUM PERENNE, Perennial rye grass.

PLATE 87.

W. Scholl. del

AGROPYRUM GLAUCUM, Colorado blue stem.

AGROPYRUM REPENS, Quack grass, Couch grass.

PLATE 89.

ELYMUS CANADENSIS, Wild rye.

PLATE 90.

ELYMUS CONDENSATUS.

PLATE 91.

ELYMUS VIRGINICUS, Wild rye.

PLATE 92.

TRIFOLIUM HYBRIDUM, Alsike clover.

PLATE 93.

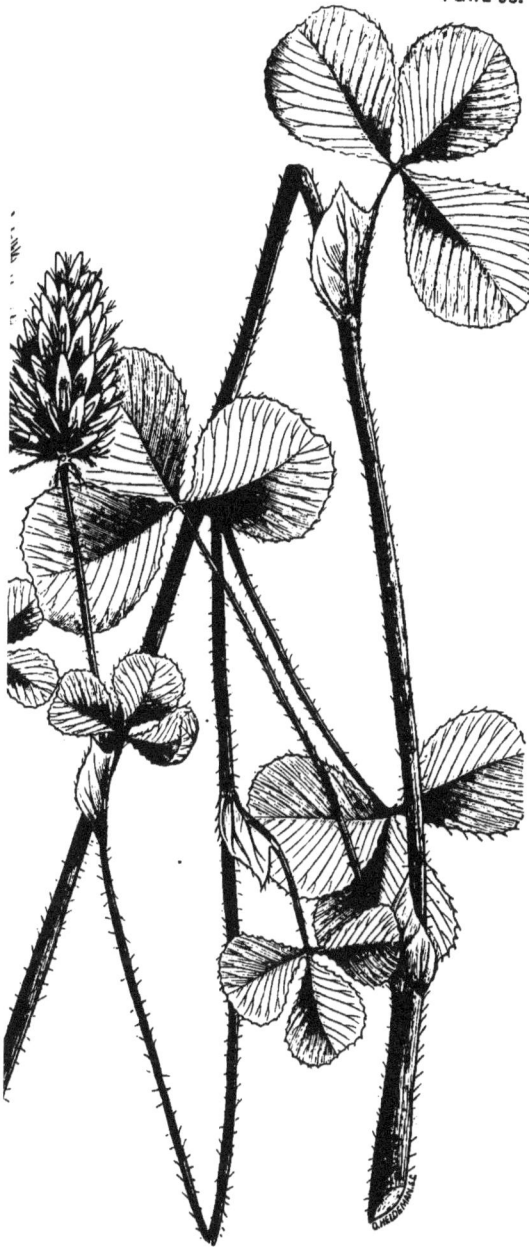

TRIFOLIUM INCARNATUM.

Marx. del.

TRIFOLIUM STOLONIFERUM, Buffalo clover.

PLATE 95.

ONOBRYCHIS SATIVA, Sainfoin, Esparsette.

PLATE 96.

Marx del

MEDICAGO SATIVA, Alfalfa, Lucerne.

PLATE 97.

W.SHOLL del

MEDICAGO DENTICULATA. Bur clover.

LESPEDEZA STRIATA, Japan clover.

PLATE 99.

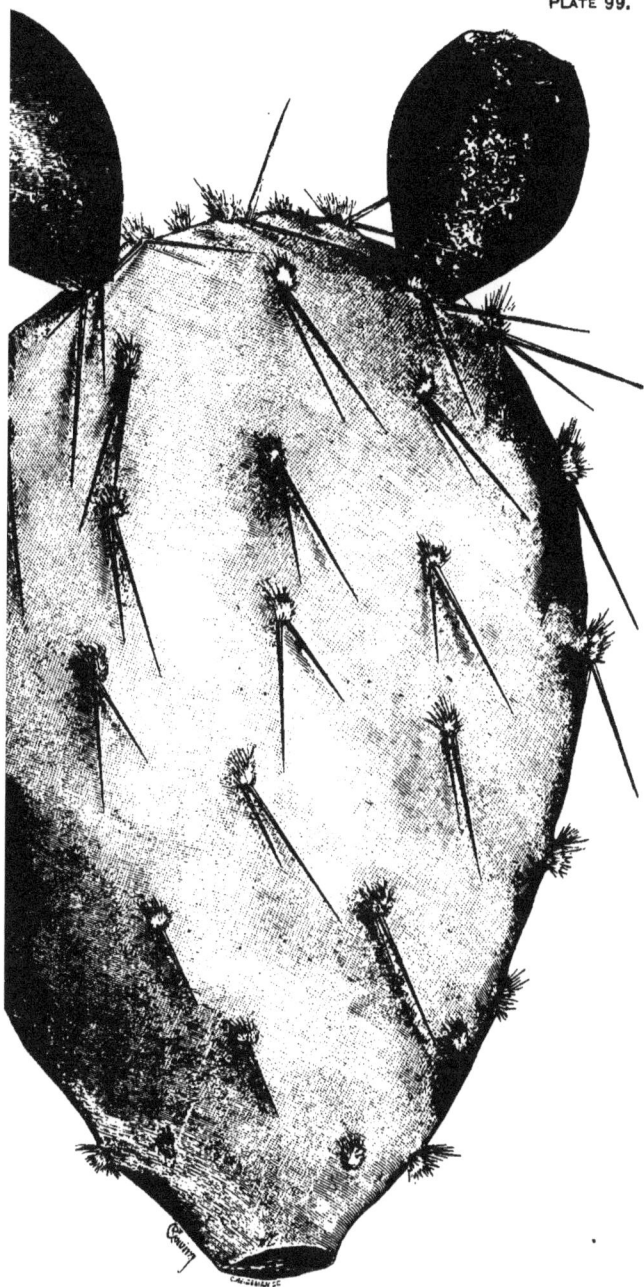

OPUNTIA ENGELMANNI, Prickly pear, Cactus.

PLATE 100.

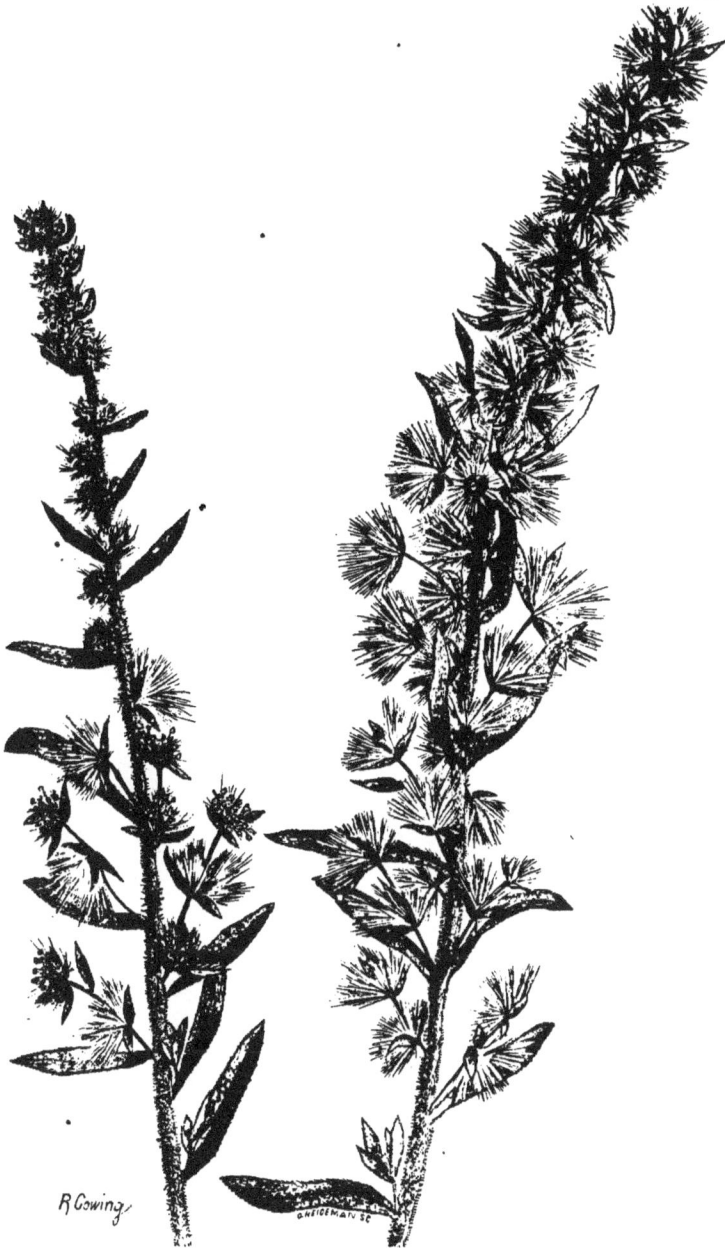

R Cowing

EUROTIA LANATA, Winter fat.

PLATE 101.

marx del

ERODIUM CICUTARIUM, Pin grass, Alfilaria.

RICHARDSONIA SCABRA, Mexican clover.

PLATE 104.

a

MABX.DEL.

H.H.NICHOLS.

MUHLENBERGIA COMATA.

PLATE 105.

AGROSTIS EXARATA var. PACIFICA.

DESCHAMPSIA CÆSPITOSA, Hair grass.

CHLORIS ALBA.

DIPLACHNE DUBIA.

PLATE 11

H.H.NICHOLS

T.TAYLOR.DEL

MELICA BULBOSA.

PLATE 11

MELICA IMPERFECTA.

PLATE 11

Marx del.

UNIOLA LATIFOLIA.

PLATE 11

GLYCERIA CANADENSIS.